SpringerBriefs in Applied Sciences and Technology

PoliMI SpringerBriefs

More information about this subseries at http://www.springer.com/series/11159
http://www.polimi.it

Alireza Entezami

Structural Health Monitoring by Time Series Analysis and Statistical Distance Measures

POLITECNICO
MILANO 1863

Alireza Entezami ⓘ
DICA
Politecnico di Milano
Milan, Italy

CED
Ferdowsi University of Mashhad
Mashhad, Iran

ISSN 2191-530X   ISSN 2191-5318   (electronic)
SpringerBriefs in Applied Sciences and Technology
ISSN 2282-2577   ISSN 2282-2585   (electronic)
PoliMI SpringerBriefs
ISBN 978-3-030-66258-5   ISBN 978-3-030-66259-2   (eBook)
https://doi.org/10.1007/978-3-030-66259-2

This Springer imprint is published by the registered company Springer Nature Switzerland AG
The registered company address is: Gewerbestrasse 11, 6330 Cham, Switzerland

Foreword

To motivate the growing research activities in the field, authors often claim that Structural Health Monitoring (SHM) is becoming a societal need in an interconnected, smart city age. On the other hand, we are flooded by news coming from the web and newspapers and reporting that structural collapses more and more often occur in developed countries, where infrastructures are older and are probably hitting, if not exceeding, their expected lifetime.

It is perceived that, in the past decades, there have been two major causes that speeded up the aging of the existing (civil) structures and infrastructures: climate change, and the increase of world population. We all know that the two causes are interrelated, as testified by the changes we are facing in our everyday life, but they also work side by side in accelerating the need for SHM strategies and deployed sensing systems to avoid catastrophic events and losses of lives and money.

In recent years, two drivers majorly pushed forward the capability of SHM strategies: the development/implementation of smart city concepts, with richness in (micro) sensors and data to be exploited for an Internet of Everything (IoE) and, specifically, to provide information hopefully linked to the heath of structural systems; the development of Artificial Intelligence-based approaches to dig into the Big Data characterizing our age. We have therefore faced a change in perspective to approach the SHM problem, from model-based methods to data-driven ones.

Terms and algorithms linked to machine and deep learning are now rather common in the field, but that does now mean that the approaches proposed in the literature are always successful, efficient so that they can even work in real time, and robust against operational and environmental variabilities. In this book, the author focuses on these specific aspects, attacking well-known problems that already became benchmarks, to test and assess the performance of the offered algorithmic analysis. Though the general approaches described in the book look simple, a wise integration is reported of learning stages, to deal with multivariate statistical datasets, and a decision-making stage, to assess the divergence of the current response from a baseline characterized by an undamaged state. As already mentioned, the presence of measurement errors and the environmental variability under continuously changing operational conditions require the SHM procedure to be extremely sensitive to (even small) damage. Pervasive sensor networks also provide rich information on the structural response to the external excitations; this introduces the issue of Big Data and the further challenge to the device and tunes effective damage-sensitive features, to be dealt with in the process of decision-making.

All the above-mentioned aspects are described and effectively considered in this book, which is therefore expected to become a good reference also for practitioners in the field on SHM.

Milan, Italy   Stefano Mariani

Preface

Structural Health Monitoring (SHM) is an important topic in civil engineering due to the great significance of civil structures such as high-rise buildings, bridges, dams, etc. The main objective of SHM is to evaluate the integrity and safety of these structures through early damage detection, damage localization, and damage quantification. Because of recent advances in sensing, data acquisition systems, and signal processing tools, data-driven methods based on statistical pattern recognition present effective and efficient algorithms for SHM by using measured vibration data. These methods are generally based on two main steps including feature extraction and statistical decision-making. Feature extraction is intended to discover meaningful information from raw measured data. Such information should be related and sensitive to structural properties and damage. In most literature of SHM, these are called damage-sensitive features. Statistical decision-making or feature classification aims at analyzing the extracted information or features in order to detect, locate, and then quantify damage by statistical approaches and machine learning algorithms. Despite valuable research studies, some challenging issues such as the negative influences of environmental and operational variations, the problem of precise damage localization, feature extraction from non-stationary signals caused by ambient vibration are still problematic and should be dealt with appropriately.

The primary motivation of this book is to propose novel methods for feature extraction and feature classification with an emphasis on addressing the major challenging issues. In this book, the fundamental principle of SHM, the theoretical background of feature extraction via time series analysis, time-frequency signal decomposition, and statistical decision-making by various univariate and multivariate distance measures are explained in detail. The book begins with an introductory chapter regarding the basic knowledge of SHM and the main concepts of statistical pattern recognition. The next chapters introduce conventional and proposed methods for feature extraction and statistical decision-making. Finally, the last two chapters present the main results and conclusions of SHM on some well-known benchmark civil structures. The author sincerely hopes that the readers will find this book interesting and useful in their research on data-driven SHM.

Milan, Italy   Alireza Entezami

Contents

Chapter 1
An Introduction to Structural Health Monitoring

1.1 Background and Motivation

Modern societies are heavily dependent upon important and complex civil engineering structures such as high-rise buildings, bridges, dams, offshore oil platforms, hospitals, nuclear power plants, railways and subways systems, etc. The other kinds of structures are heritage buildings, ancient bridges, and holy shrines that are identities of a society or a culture. Many of these existing structural systems are currently nearing the end of their original design life due to operational loading, environmental conditions, and being aging. Furthermore, some of them are vital structures (e.g. nuclear power plants) that may seriously threaten human health in case of any deterioration. Sometimes there are no engineering and economic justifications to reconstruct structures such as heritage buildings, ancient bridges, and holy shrines. For these reasons, it is indispensable to evaluate the states of civil engineering structures in terms of structural condition assessment and damage detection in order to increase structural safety, enhance their performances and serviceability, reduce retrofit and rehabilitation costs, provide a proper operation, and prevent any irreparable event caused by damage and deterioration.

Structural health monitoring (SHM) is an active and practical research area in civil, engineering societies that essentially aims to assess the health and integrity of structures by monitoring measured vibration data and detecting any probable damage by various vibration-based techniques (Farrar and Worden 2013). This process can be carried out in the mechanical and aerospace engineering with the name of condition monitoring (CM). It is analogous to SHM but specifically addresses fault detection in mechanical and aerospace systems such as rotating machinery, gearbox, wind turbine blades, and systems, etc. (Worden et al. 2015).

In the most general terms, the damage is defined as either intentionally or unintentionally adverse changes in a structure that significantly affect the current or future performance of that system, which can be either natural or man-made. The occurrence of a damage scenario in a structure may be stemming from undesirable changes

A. Entezami, *Structural Health Monitoring by Time Series Analysis and Statistical Distance Measures*, PoliMI SpringerBriefs, https://doi.org/10.1007/978-3-030-66259-2_1

in geometry configurations, boundary conditions, and materials that lead to cracks in concrete, loose bolts and broken welds in steel connections, corrosion, fatigue, etc. These may cause permanent changes in structural stiffness, inappropriate stresses and displacements, unfavorable vibrations, adverse dynamic behavior, failure, and even collapse. Having considered the above-mentioned descriptions, this chapter of the book is intended to provide a general overview of SHM and its main objectives for civil engineering, present the general SHM methods, and introduce the application of statistical pattern recognition to SHM.

1.2 Levels of SHM

The main aim of an SHM strategy is to monitor a structure and immediately alarm the occurrence of damage. Further to this, the other important objectives of SHM are to identify the location of damage and then estimate the level of damage severity. These were initially described by Rytter (1993). Accordingly, an SHM procedure can be classified into damage diagnosis and damage prognosis. The process of damage diagnosis is decomposed into three levels including:

- Level 1—Early damage detection.
- Level 2—Damage localization.
- Level 3—Damage quantification.

The first level is a global strategy aiming at evaluating whether the damage has occurred throughout the structure (Is there damage in the system?). By contrast, the second and third levels are local strategies. In this regard, Level 2 is intended to identify the location of damage (Where is the damage in the system?) after detecting damage based on Level 1. Further to early damage detection and localization, it is attempted to estimate or quantify the severity of damage (How severe is the damage?). As the level increases, the knowledge about damage and the complexity of SHM methods increase as well (Tondreau and Deraemaeker 2014). On this basis, the simplest way is the early damage detection due to the applicability of global vibration measurements (e.g. modal frequencies) and simple formulations (e.g. relative errors in the modal frequencies of a bridge in different conditions). In order to locate damage, it may be difficult to utilize global vibration data. Moreover, this process is more difficult in large and complex civil structures. Determination of the sensor placement for acquiring sufficient vibration information is a prominent task of the level of damage localization (Entezami and Shariatmadar 2019; Papadimitriou et al. 2000; Entezami et al. 2020f; Capellari et al. 2018). For both localizing and quantifying damage, it is important to have knowledge about the damaged condition of the structure (Biondini et al. 2008; Biondini and Restelli 2008). This procedure may be impossible because sometimes there are no engineering and economic justifications to induce intentional damage to the real structure in order to establish a damage quantification framework. This limitation is normally dealt with by finite element model updating, in which case it is assumed that the accurate finite element (FE) model of the real structure refers to the healthy or normal condition (Sarmadi et al. 2016; Entezami et al. 2017; Rezaiee-Pajand et al. 2019). Therefore, one can apply diverse damage scenarios to the FE model of the structure in order to quantify the severity of the damage.

Eventually, the damage prognosis is defined as the estimation and/or prediction of the remaining service life of the structure (How much useful life remains?). This estimate is based on a quantified definition of the structure failure, an assessment of the structure's current damaged state, and the output of models that predict the propagation of the damage in the structure based on some estimates of the structure's future loading. In other words, the process of damage prognosis attempts to forecast the performance of the structure by assessing its current state (i.e. SHM), estimating the future loading environments, and predicting through simulation and experience for the remaining useful life of the structure. Such remaining life predictions will necessarily rely on information from usage monitoring; SHM; past, current and anticipated future environmental and operational conditions; the original design assumptions regarding loading and operational environments, previous component, and system-level testing; and any available maintenance records (Farrar and Worden 2013, Chap. 14).

1.3   Methods of SHM

The levels of damage diagnosis are normally carried out by model-driven or data-driven methods (Barthorpe 2010). The main premise behind the first approach is to use an elaborate and accurate analytical or FE model, while the second method is only based on the application of vibration responses without any FE modeling.

1.3.1   Model-Driven Approaches

A model-driven method is usually based on the development of an FE model of a real structure and the use of the main concepts of finite element model updating (Sarmadi et al. 2016; Eftekhar Azam et al. 2017; Eftekhar Azam and Mariani 2018; Biondini and Vergani 2012; Biondini 2004; Papadimitriou et al. 1997; Entezami et al. 2014; Entezami et al. 2020b). Accordingly, one assumes that the FE and experimental models serve as normal and damaged conditions of the real structure (Entezami et al. 2017; Entezami et al. 2020c). Having considered analytical data associated with the FE model and the measured vibration data obtained from dynamic tests implemented on the real structure, it is possible to detect, locate, and quantify damage by updating the main characteristics of the FE model, which are pertinent to damage such as the structural stiffness (Pedram et al. 2017). In fact, the procedures of damage detection, localization, and quantification can be performed by the solution of an inverse problem using experimental data collected from the structure under test.

The main limitation of the model-driven approaches is that the FE model of the structure, which behaves like a normal or undamaged condition, differs from its corresponding experimental model. In such a case, it is necessary to initially adjust the FE model or normal condition of the real structure by model updating techniques (Mottershead et al. 2011; Sehgal and Kumar 2016; Moaveni et al. 2009; Moaveni et al. 2008; Papadimitriou and Ntotsios 2009; Giagopoulos et al. 2019; Sarmadi et al. 2020b; Rezaiee-Pajand et al. 2020). Another limitation in applying model-driven methods is concerned with the ill-posedness of the inverse problem. To deal with this issue, one needs to utilize robust regularized solution techniques such Tikhonov regularization (Zhang and Xu 2016; Entezami and Shariatmadar 2014), truncated singular value decomposition (Weber et al. 2009), RLMSR (Entezami et al. 2017), etc., which essentially require obtaining appropriate regularization parameters.

1.3.2   Data-Driven Approaches

A data-driven method is an alternative way to implement an SHM strategy. This method involves the construction of a statistical model rather than the FE model of the real structure. The model of interest is established by applying statistical pattern recognition to experimentally measured vibration data associated with the normal and damaged conditions of the real structure. Since the data-driven approaches rely entirely upon the data, one can avoid the construction of the FE model of the structure. Furthermore, the main advantage of the data-driven methods is the ability to incorporate various uncertainties arising from measurement variability, noise, operational and environmental conditions into the process of SHM. Taking the limitations and drawbacks of the model-driven techniques into consideration, it seems that the data-driven methods are more suitable for SHM applications (Farrar and Worden 2013).

1.4   Statistical Pattern Recognition

Statistical pattern recognition is related to the use of statistical techniques for analyzing data measurements in order to extract meaningful information and make justified decisions (Webb and Copsey 2011). In the context of SHM, the statistical pattern recognition involves an initial evaluation of the necessity of implementing SHM on different structural systems (operational evaluation), the acquisition of vibration data from sensors (sensing and data acquisition), the extraction of meaningful information or damage-sensitive features from raw vibration measurements, and the statistical analysis or decision-making for feature classification (Farrar and Worden 2013). The great advantage of the statistical pattern recognition is the possibility of implementing an automatic SHM process, which enables to remove the need for the intervention of human experts as far as possible (Entezami et al. 2020a; Entezami et al. 2020d).

1.4.1 Operational Evaluation

The process of operational evaluation attempts to provide answers to some important questions regarding the implementation of an SHM strategy on a structure such as:

- What are the life-safety and/or economic justifications for performing the SHM?
- How is damage defined for the system being investigated and, for multiple damage possibilities?
- What are the conditions, both operational and environmental, under which the system to be monitored functions?
- What are the limitations of acquiring data in the operational environment?

In some cases, a significant effort will have to be expended to provide quantified answers to these questions. However, without well-defined answers to these questions, it is unlikely that SHM systems will be developed or be used for anything but research programs. For example in the part of economic justification, numerous SHM studies have been carried out with little concern for the economic issues associated with field implementation of the technology. This situation has arisen because many of these studies have been carried out as part of a more basic, one-of-a-kind, research effort aimed at developing proof-of-concept technology demonstrations. However, it is unlikely that any private company or government institution will be willing to fund the development of an SHM capability for field deployment unless it can be shown that this capability can provide a quantifiable and enhanced life-safety and/or economic benefit relative to the currently employed damage detection strategy. Economic benefits can be realized in a variety of manners including reduced maintenance cycles, reduced warranty obligations, increased manufacturing capacity, and increased system availability.

The process of demonstrating the economic benefits of a damage detection capability will set limits on all other aspects of the SHM paradigm. In particular, the budget of data acquisition hardware and the budget for the time that can be spent developing numerical simulations of the damage scenarios will be limited by the need to show a positive rate-of-return on investment in SHM technology. Therefore, if one is to take SHM beyond the proof-of-concept level, at a minimum, the following questions related to the life-safety/economic benefits issue must be presented to authorities deciding to invest in the system:

- What are the limitations of the currently employed damage detection methodology?
- What are the advantages provided by the proposed SHM system?
- How much will the proposed SHM system cost?

A final economic issue that should be considered is the portion of the budget that is funding the SHM system development. Typically, for most high-expenditure aerospace, civil and mechanical systems, the design and construction budget is much larger than the annual maintenance budget. Therefore, the capital expenditures associated with the hardware requirements of the SHM system will not seem as extreme if they are included in the design and construction portion of the budget. In contrast, this same hardware can be a significant fraction of the system's annual maintenance budget, which poses challenging economic issues when trying to incorporate an SHM capability in a retrofit mode (Farrar and Worden 2013, Chap. 3).

1.4.2   Sensing and Data Acquisition

The acquisition of accurate measurements from the dynamic response of a structure is essential to SHM. The part of sensing and data acquisition in SHM involves the selection of excitation methods and sensor types, the specification of sensor numbers and locations, and the choice of data acquisition/storage/transmittal hardware. However, economic considerations play major roles in making decisions about the selection of sensors and data acquisition systems (Farrar and Worden 2013; Wang et al. 2014; Capellari et al. 2017, 2018).

Almost all sensor systems are deployed for one or more applications including (i) detection and tracking problems, (ii) model development, validation and uncertainty quantification, and (iii) control systems. Because SHM necessarily focuses on the problems of detection and tracking, the first goal of any sensor system development is to make the sensor readings, which should directly be correlated with and sensitive to damage (i.e. this concept refers to the detection problem). Subsequently, it is attempted to acquire sensor readings and information extracted from these data that enable us to observe changes in measurements attributable to increasing damage levels (i.e. this concept refers to the tracking problem). There are a variety of approaches to developing a sensing and data acquisition strategy for achieving the SHM goals including (Farrar and Worden 2013, Chap. 4):

- **Approach I**: An often-used approach to deploying a sensing system for damage detection, particularly associated with earlier SHM studies, consists of adopting a sparse sensor array. This array is installed on the structure after fabrication, possibly following an extended period of service. Sensors are typically chosen based on previous experience of the SHM system developer and commercial availability. The selected sensing systems often have been commercially available for some time and the technology may be more than twenty years old. Excitation is often limited to that provided by the ambient operational environment. The physical quantities that are measured are often selected in an ad hoc manner without an a priori quantified definition of the damage that is to be detected or any a priori analysis that would indicate that these measured quantities are sensitive to the damage of interest. This approach dictates that damage-sensitive data features are selected 'after the fact' using archived sensor data and ad hoc algorithms. The most common damage detection methods associated with *Approach I* are based on the assumption that the undamaged and damaged conditions of the structure are subjected to nominally similar excitations (Farrar and Worden 2013, Chap. 4).

- **Approach II**: This is a more coupled analytical/experimental approach to the sensor system definition that incorporates some significant improvements over *Approach I*. First, the damage is well defined and to some extent quantified through the operational evaluation process before the sensing system is designed. Next, the sensing system properties and any relevant actuator properties are defined based on the results of numerical simulations of the dynamic responses of the damaged structure or physical experiments. The data analysis procedures that will be employed in the damage detection application are also considered when developing the sensing and data acquisition system. This process of defining the sensor system properties will often be iterative. Sensor types and locations are chosen because the numerical simulations or physical tests and associated optimization procedures show that the expected type of damage produces known, observable, and statistically significant effects in features derived from the measurements at these locations. Additional sensing requirements are then defined based on how changing operational and environmental conditions affect the damage detection process. Methods for data archiving and telemetry are considered in the sensor system design along with long-term environmental ruggedness and maintenance issues. However, all sensors and data acquisition hardware are still chosen from the commercially available products that best match the defined sensing system requirements.

The second approach incorporates several enhancements that would typically improve the probability of damage diagnosis including:

- It has well-defined and quantified damage information that is based on initial system design information, numerical simulation of the postulated damage process, qualification of test results, and maintenance records.
- It uses sensors that are shown to be sensitive enough to provide data that can be used to identify the predefined damage when the measured data are coupled with the feature extraction and statistical modeling procedures.
- Active sensing is incorporated into the process where a known input is used to excite the structure with a waveform tailored to the damage detection process.
- Sensors are placed at locations where responses are known from analysis, experiments, and past experience to be sensitive to damage.
- Additional measurements are made that can be used to quantify changing operational and environmental conditions.

The other important issue in the portion of data acquisition is the type of data for SHM. Instrumentation, which includes sensors and data acquisition hardware, first translates the vibration response of the structure into an analog voltage signal that is proportional to the response quantity of interest. This process is typically accomplished through the transduction properties of the sensing material that allows for the conversion of the response variable (e.g. acceleration) to some other field (most often an electric signal). Subsequently, the analog signal is discretely sampled to produce digital data. Most of the vibration data used in the process of SHM include (Farrar and Worden 2013):

- Dynamic input and response quantities (e.g. input force, strain or acceleration).
- Physical quantities (e.g. electromagnetic fields, chemicals).
- Environmental quantities (e.g. temperature or wind speed).
- Operational quantities (e.g. traffic volume or vehicle airspeed).

Sensing systems for SHM consist of some or all of the following components (Wang et al. 2014):

- Transducers that convert changes in the field variable of interest (e.g. acceleration, strain, temperature) to changes in an electrical signal (e.g. voltage, impedance, resistance)
- Actuators that can be used to apply a prescribed input to the system (e.g. a piezoelectric transducer bonded to the surface of a structure).
- Analog-to-digital (A/D) converters that transform the analog electrical signal into a digital signal (i.e. this signal can subsequently be processed on digital hardware). For cases where actuators are used, a digital-to-analog (D/A) converter will also be needed to change a prescribed digital excitation signal to an analog voltage that can be used to control the actuator).
- Signal conditioning.
- Power.
- Telemetry.
- Signal processing.
- Memory for data storage.

The number of sensing systems available for SHM even now is enormous and these systems vary depending upon the specific SHM activity. However, wired and wireless sensors are two common types of sensing systems. The wired SHM systems are defined as ones that transfer data and power to or from the sensor over a direct-wired connection from the transducer to the central data analysis facility. In contrast to this type of sensing system, the application of wireless sensors has been considerably increased over the wired sensing system. Sensing and data acquisition systems are broad fields of SHM that are beyond the scope of this research work. For more details, the reader can be referred to Farrar and Worden (2013) and Wang et al. (2014).

1.4.3 Feature Extraction

Feature extraction refers to the process of discovering meaningful information (pattern or features) from the raw vibration measurements (Farrar and Worden 2013, Chaps. 7 and 8). These features are often known as damage-sensitive features (DSFs). Dynamic and/or statistical characteristics extracted from the vibration data are the most important DSFs that are used in the context of SHM. The main characteristic of such features is that both of them should be pertinent to damage, which means that the correlation between the DSFs and damage is more readily observable. The extraction of DSFs that can accurately distinguish a damaged structure from an undamaged one is the primary and significant topic in the context of SHM based on statistical pattern recognition. For this purpose, advanced signal processing techniques in time, frequency, and time–frequency domains provide reliable and diverse approaches to feature extraction (Amezquita-Sanchez and Adeli 2016; Sarmadi et al. 2020a). In fact, signal processing is the key component of feature extraction for data-driven SHM techniques. Feature extraction by signal processing is the most challenging issue in SHM due to the complex processes involved in the structural response to dynamic loading. A majority of dynamic characteristics that are suitable to use as the DSFs are (Farrar and Worden 2013):

- Impulse and frequency response functions.
- Coherence function.
- Power spectral and cross-spectral density functions.
- Modal parameters and functions (e.g. modal frequencies, mode shapes, modal strain energy, modal flexibility, mode shape curvature, etc.).
- Fast Fourier transform.
- Wavelet transform.
- Hilbert-Huang transform.

In contrast, the statistical features originate from the statistical concepts and signal processing techniques in time, frequency, and time–frequency domains such as (Farrar and Worden 2013; Amezquita-Sanchez and Adeli 2016):

- Statistical moments such as the mean, standard deviation, skewness, kurtosis, etc.
- Peak amplitudes of the vibration signal.
- Root-mean-square.
- Crest factor and K-factor.
- Probability distributions.
- Time series analysis.

1.4.4   Statistical Decision-Making

Statistical decision-making is concerned with the implementation of algorithms that operate on the DSFs to analyze them for damage diagnosis and prognosis. After the feature extraction, statistical decision-making is a strategy for feature classification or discrimination. In the context of SHM, the feature classification refers to the levels of early damage detection, damage localization, and damage quantification. Because the functional relationship between the DSFs and damage is often difficult to define based on physics-based engineering analysis procedures, statistical approaches are derived by using the statistical pattern recognition paradigm. For SHM applications, the main idea behind the statistical decision-making is to learn a statistical model or classifier by some available DSFs (i.e. training data) and make decisions about the status of new DSFs (i.e. testing data) (Farrar and Worden 2013, Chap. 9).

Despite the importance of statistical decision-making for SHM, the accuracy of damage diagnosis and prognosis highly pertains to the sensitivity of DSFs. In the case of extracting reliable and accurate features, statistical decision-making can simply indicate any adverse changes in the structure. This means that if the extracted and selected features are not correlated with the damage, even the most robust and powerful statistical approaches fail in detecting damage.

1.5   Machine Learning

The process of feature classification in statistical decision-making is usually implemented under the theory of machine learning. In general, there are two classes of machine learning in the area of SHM including supervised learning and unsupervised learning (Farrar and Worden 2013). For the applications of SHM, the supervised learning class aims to train a classifier or statistical model by using the DSFs extracted from both the undamaged and damaged conditions (Farrar and Worden 2013, Chap. 11), while the unsupervised learning class only needs the DSFs of the undamaged state for learning the classifier of interest (Farrar and Worden 2013, Chap. 10). The fundamental axiom of SHM lies in the fact that the damage diagnosis is based on the comparison between two structural states including a normal (known) condition and a current (unknown) state, which can be undamaged or damaged (Worden et al. 2007). Accordingly, the use of a supervised learning algorithm for training a classifier encounters a major obstacle associated with the unavailability of the DSFs of the unknown (damaged) state. On the other hand, it is not reasonable and economical to induce intentional damage to the large-scale and vital civil structures in an effort to provide information about their damaged states. Because the unsupervised learning algorithm does not need any information (i.e. vibration data or DSFs) of the structure in any damaged condition to learn the classifier, the unsupervised learning class is more beneficial to SHM compared with the supervised learning (Entezami et al. 2020g; Sarmadi and Entezami 2020). The process of statistical decision-making based on the unsupervised learning class is carried out under two main phases including training or baseline and monitoring or inspection phases (Fassois and Sakellariou 2009).

During the training or baseline period, the DSFs of the undamaged or normal condition of the structure are extracted to construct the training data and learn a statistical model. In the monitoring or inspection phase, new DSFs of the current state of the structure are extracted and utilized in the trained model as the testing data to make a decision in terms of damage detection, localization, and quantification. Therefore, the process of statistical decision-making for damage diagnosis in the monitoring state is implemented by comparing the DSFs of the undamaged and damaged conditions based on the learned model in the training phase. Any deviation of testing data from the trained model is indicative of the occurrence of damage.

1.6 Environmental and Operational Variability

Many vibration-based techniques have been proposed to detect, locate, and quantify damage by using changes in the dynamic characteristics of structures. It is well known that structural damage will cause changes in measured vibration data of structures. However, the same consequence is obtainable when those are subjected to environmental (i.e. wind, temperature, and humidity) and operational (i.e. ambient loading conditions, operational speed, and mass loading) variability conditions (Sohn 2007). For example, the environmental variability caused by temperature fluctuations not only changes the material stiffness but also alters the boundary conditions of a structure. In such a case, it is expected that the stiffness of the structure changes leading to variations in dynamic responses such as natural frequencies. The major challenging issue regarding environmental and operational variability (EOV) is that such conditions exhibit the same behavior as damage (Kullaa 2011). Under such circumstances, the data-driven techniques may suffer from false-positive or Type I errors (i.e. the structure is undamaged but the method alarms damage) and false-negative or Type II errors (i.e. the structure is damaged but the method declares the undamaged state) (Sarmadi and Karamodin 2020; Entezami et al. 2020e; Sarmadi et al. 2020c).

1.7 Aim and Scope of the Book

In order to prevent terrible events caused by the occurrence of damage and decrease the high costs of maintenance and rehabilitation in the important and expensive civil structures, SHM is a necessity for every society, regardless of culture, geographical location, and economic development. Due to the great importance of SHM and superiority of the data-driven method over the model-driven approach, the main objective of this book is to conduct effective and efficient research on the data-driven SHM strategy through the statistical pattern recognition in terms of the feature extraction and statistical decision-making for the damage diagnosis under the EOV conditions.

Many innovative and effective feature extraction methods by time series analysis and signal processing are proposed to extract reliable DSFs from vibration time-domain responses. In this regard, some limitations and drawbacks of time series modeling are dealt with. For statistical decision-making, the focus of this research work is on distance-based techniques. Accordingly, novel methods under the unsupervised learning manner are proposed to detect, locate, and quantify different damage scenarios.

1.8 Organization of This Book

The organization of the book is as follows. Chapter 2 presents the methods of feature extraction via time series analysis under stationary vibration responses. For nonstationary data, Chap. 3 proposes some hybrid methods as combinations of time–frequency data analysis techniques and time-invariant linear models. Distance-based methods for damage diagnosis are proposed in Chap. 4. Chapter 5 presents the results of damage diagnosis on several experimental structures based on the proposed methods and some comparative analyses. Eventually, the conclusions of this research work are stated in Chap. 6.

1.9 Conclusions

This introductory chapter has presented the definitions and objectives of SHM and statistical pattern recognition, the levels of SHM, the general SHM methods, and the effects of EOV conditions. From this chapter of the book, one can conclude that SHM is a great necessity for the guarantee of the safety and integrity of significant civil structures. To achieve this objective, it is necessary to implement the three main steps of the damage diagnosis including early damage detection, damage localization, and damage quantification. The data-driven methods are more applicable to SHM compared with the model-driven techniques due to some advantages such as the lack of modeling the finite element model of the structure and implementing the model updating strategies. For data-driven methods, the procedures of feature extraction and statistical decision-making are of paramount importance to SHM. Therefore, the main focus of this research work is on these procedures. Since the EOV conditions are available in most cases of SHM applications, it is essential to pay more attention to their effects and avoid false alarm and false detection errors.

References

Amezquita-Sanchez JP, Adeli H (2016) Signal processing techniques for vibration-based health monitoring of smart structures. Arch Comput Methods Eng 23(1):1–15

Barthorpe RJ (2010) On model-and data-based approaches to structural health monitoring. Doctoral dissertation, University of Sheffield

Biondini F (2004) A three-dimensional finite beam element for multiscale damage measure and seismic analysis of concrete structures. In: 13th world conference on earthquake engineering, August 1–6, 2004, Vancouver, British Columbia, Canada, 13 WCEE Secretariat, Paper No. 2963. Paper No. 2963, Vancouver, BC, Canada, pp 1–6

Biondini F, Restelli S (2008) Measure of structural robustness under damage propagation. In: Fourth international conference on bridge maintenance, safety, and management (IABMAS 2008), July 13–17, 2008, Seoul, Korea, CRC Press, Taylor & Francis Group, pp 1–8

Biondini F, Vergani M (2012) Damage modeling and nonlinear analysis of concrete bridges under corrosion. In: Sixth international conference of bridge maintenance, safety and management (IABMAS 2012), 8–12 July, 2012, Lake Como, Italy, CRC Press, Taylor & Francis Group, pp 949–957

Biondini F, Frangopol DM, Garavaglia E (2008) Damage modeling and life-cycle reliability analysis of aging bridges. In: Fourth international conference on bridge maintenance, safety, and management (IABMAS 2008), July 13–17, 2008, Seoul, Korea, CRC Press, Taylor & Francis Group, p 452

Capellari G, Chatzi E, Mariani S (2017) Cost-benefit optimization of sensor networks for SHM applications. In: multidisciplinary digital publishing institute proceedings, vol 3, p 132

Capellari G, Chatzi E, Mariani S (2018) Structural health monitoring sensor network optimization through bayesian experimental design. ASCE-ASME J Risk Uncertainty Eng Syst, Part A: Civ Eng 4(2):04018016

Eftekhar Azam S, Mariani S (2018) Online damage detection in structural systems via dynamic inverse analysis: a recursive Bayesian approach. Eng Struct 159:28–45

Eftekhar Azam S, Mariani S, Attari KAN (2017) Online damage detection via a synergy of proper orthogonal decomposition and recursive Bayesian filters. Nonlinear Dyn 89(2):1489–1511

Entezami A, Shariatmadar H (2014) Damage detection in structural systems by improved sensitivity of modal strain energy and Tikhonov regularization method. Int J Dyn Control 2(4):509–520

Entezami A, Shariatmadar H (2019) Damage localization under ambient excitations and non-stationary vibration signals by a new hybrid algorithm for feature extraction and multivariate distance correlation methods. Struct Health Monit 18(2):347–375

Entezami A, Shariatmadar H, Ghalehnovi M (2014) Damage detection by updating structural models based on linear objective functions. J Civ Struct Health Monit 4(3):165–176

Entezami A, Shariatmadar H, Sarmadi H (2017) Structural damage detection by a new iterative regularization method and an improved sensitivity function. J Sound Vib 399:285–307

Entezami A, Shariatmadar H, Sarmadi H (2020) Condition assessment of civil structures for structural health monitoring using supervised learning classification methods. Iran J Sci Technol, Trans Civ Eng 44(1):51–66

Entezami A, Shariatmadar H, Mariani S (2020) Early damage assessment in large-scale structures by innovative statistical pattern recognition methods based on time series modeling and novelty detection. Adv Eng Softw 150:102923

Entezami A, Shariatmadar H, Mariani S (2020) Fast unsupervised learning methods for structural health monitoring with large vibration data from dense sensor networks. Struct Health Monit 19(6):1685–1710

Entezami A, Sarmadi H, Behkamal B, Mariani S (2020) Big data analytics and structural health monitoring: a statistical pattern recognition-based approach. Sensors 20(8):2328

Entezami A, Sarmadi H, Razavi BS (2020) An innovative hybrid strategy for structural health monitoring by modal flexibility and clustering methods. J Civ Struct Health Monit 10(5):845–859

Entezami A, Sarmadi H, Saeedi Razavi B (2020c) An innovative hybrid strategy for structural health monitoring by modal flexibility and clustering methods. J Civ Struct Health Monit, In press

Entezami A, Sarmadi H, Salar M, De Michele C, Arslan AN (2020d) A novel data-driven method for structural health monitoring under ambient vibration and high-dimensional features by robust multidimensional scaling. Structural Health Monitoring, in press

Farrar CR, Worden K (2013) Structural health monitoring: a machine learning perspective. Wiley

Fassois SD, Sakellariou JS (2009) Statistical time series methods for structural health monitoring. In: Encyclopedia of structural health monitoring. Wiley, pp 443–472

Giagopoulos D, Arailopoulos A, Dertimanis V, Papadimitriou C, Chatzi E, Grompanopoulos K (2019) Structural health monitoring and fatigue damage estimation using vibration measurements and finite element model updating. Struct Health Monit 18(4):1189–1206

Kullaa J (2011) Distinguishing between sensor fault, structural damage, and environmental or operational effects in structural health monitoring. Mech Syst Signal Process 25(8):2976–2989

Moaveni B, He X, Conte JP, De Callafon RA (2008) Damage identification of a composite beam using finite element model updating. Comput-Aided Civ Infrastruct Eng 23(5):339–359

Moaveni B, Conte JP, Hemez FM (2009) Uncertainty and sensitivity analysis of damage identification results obtained using finite element model updating. Comput-Aided Civ Infrastruct Eng 24(5):320–334

Mottershead JE, Link M, Friswell MI (2011) The sensitivity method in finite element model updating: a tutorial. Mech Syst Signal Process 25(7):2275–2296

Papadimitriou C, Ntotsios E Structural model updating using vibration measurements. In: ECCOMAS thematic conference on computational methods in structural dynamics and earthquake engineering, June 22–24, 2009, Rhodes, Greece

Papadimitriou C, Levine-West M, Milman M (1997) Structural damage detection using modal test data. Struct Health Monit-Curr Status Perspect:678–689

Papadimitriou C, Beck JL, Au S-K (2000) Entropy-based optimal sensor location for structural model updating. J Vib Control 6(5):781–800

Pedram M, Esfandiari A, Khedmati MR (2017) Damage detection by a FE model updating method using power spectral density: numerical and experimental investigation. J Sound Vib 397:51–76

Rezaiee-Pajand M, Entezami A, Sarmadi H (2019) A sensitivity-based finite element model updating based on unconstrained optimization problem and regularized solution methods. Struct Control Health Monit 27(5):1–29

Rezaiee-Pajand M, Sarmadi H, Entezami A (2020) A hybrid sensitivity function and Lanczos bidiagonalization-Tikhonov method for structural model updating: Application to a full-scale bridge structure. Appl Math Model 89:860–884

Rytter A (1993) Vibrational based inspection of civil engineering structures. University of Aalborg, Aalborg, Denmark

Sarmadi H, Entezami A (2020) Application of supervised learning to validation of damage detection. Arch Appl Mech

Sarmadi H, Karamodin A (2020) A novel anomaly detection method based on adaptive Mahalanobis-squared distance and one-class kNN rule for structural health monitoring under environmental effects. Mech Syst Sig Process 140:106495

Sarmadi H, Karamodin A, Entezami A (2016) A new iterative model updating technique based on least squares minimal residual method using measured modal data. Appl Math Model 40(23–24):10323–10341

Sarmadi H, Entezami A, Ghalehnovi M (2020) On model-based damage detection by an enhanced sensitivity function of modal flexibility and LSMR-Tikhonov method under incomplete noisy modal data. Eng Comput. https://doi.org/10.1007/s00366-020-01041-8

Sarmadi H, Entezami A, Daneshvar Khorram M (2020) Energy-based damage localization under ambient vibration and non-stationary signals by ensemble empirical mode decomposition and Mahalanobis-squared distance. J Vib Control 26(11–12):1012–1027

Sarmadi H, Entezami A, Saeedi Razavi B, Yuen K-V (2020c) Ensemble learning-based structural health monitoring by Mahalanobis distance metrics. Struct Control Health Monit:e2663

Sehgal S, Kumar H (2016) Structural dynamic model updating techniques: a state of the art review. Arch Comput Methods Eng 23(3):515–533

Sohn H (2007) Effects of environmental and operational variability on structural health monitoring. Philos Trans Royal Soc London A: Math, Phys Eng Sci 365(1851):539–560

Tondreau G, Deraemaeker A (2014) Numerical and experimental analysis of uncertainty on modal parameters estimated with the stochastic subspace method. J Sound Vib 333(18):4376–4401

Wang ML, Lynch JP, Sohn H (2014) Sensor technologies for civil infrastructures: applications in structural health monitoring. Woodhead Publishing (Elsevier)

Webb AR, Copsey KD (2011) Statistical pattern recognition, 3rd edn. Wiley

Weber B, Paultre P, Proulx J (2009) Consistent regularization of nonlinear model updating for damage identification. Mech Syst Sig Process 23(6):1965–1985

Worden K, Farrar C, Manson G, Park G (2007) The fundamental axioms of structural health monitoring. In: Proceedings of the royal society a: mathematical, physical and engineering science

Worden K, Cross EJ, Dervilis N, Papatheou E, Antoniadou I (2015) Structural health monitoring: from structures to systems-of-systems. IFAC-Pap Online 48(21):1–17

Zhang CD, Xu YL (2016) Comparative studies on damage identification with Tikhonov regularization and sparse regularization. Struct Control Health Monit 23(3):560–579

Chapter 2
Feature Extraction in Time Domain for Stationary Data

2.1 Introduction

Time series analysis is an efficient and powerful statistical tool for signal processing and feature extraction. In most cases, a time series model is fitted to the vibration time-domain measurements and some statistical properties such as the model coefficients and residuals are extracted as the DSFs (Fassois and Sakellariou 2009). When the vibration measurements are linear and stationary, the most effective approach is to use time-invariant linear models including AutoRegressive (AR), AutoRegressive with eXogenous input (ARX), AutoRegressive Moving Average (ARMA), AutoRegressive Moving Average with eXogenous input (ARMAX), and AutoRegressive-AutoRegressive with eXogenous input (ARARX). More details can be found in Chap. 3 of this research work and Ljung (1999). The application of the coefficients of the AR model as the DSFs to SHM can be found in (Sohn et al. 2000; de Lautour and Omenzetter 2010; Gul and Necati Catbas 2009; Datteo and Lucà 2017; Entezami et al. 2019a; Entezami et al. 2020e; Sarmadi and Entezami 2020). Detectability of damage by using the AR model coefficients was evaluated by Hoell and Omenzetter (2016), who presented an optimal selection of such DSFs for early damage detection on wind turbine blades. Furthermore, Yao and Pakzad (2013) evaluated the sensitivity of the coefficient of the AR model to both structural damage and noise.

One of the earliest research articles regarding the use of the AR model residuals as the DSFs in SHM is related to Fugate et al. (2001). Mattson and Pandit (2006) exploited the statistical moments of the residuals of the Vector AutoRegressive (VAR) model such as the mean, standard deviation, Skewness, and Kurtosis in an effort to locate damage. Some authors also utilized both the coefficients and residuals of the AR model in SHM (Zheng and Mita 2009; Yao and Pakzad 2012; Rezaiee-Pajand et al. 2017; Entezami et al. 2019a; Entezami and Shariatmadar 2017, 2019b; Entezami et al. 2018a, b; Entezami et al. 2020c).

In the case of using ambient vibration as the excitation source applied to structures, SHM is an output-only process in the sense that the excitation forces are unknown and immeasurable. Under such circumstances, Carden and Brownjohn (2008) concluded that the ambient vibration directly affects the error or Moving Average (MA) term of time series models. The same conclusion was obtained by Yao and Pakzad (2013), who mentioned that the vibration responses acquired from the ambient excitation sources conform to the ARMA model. Nair et al. (2006) used this model for damage detection and localization in the ASCE benchmark structure (Johnson et al. 2004; Dyke et al. 2003) by proposing two damage localization indices using the first three AR coefficients of the ARMA representation. The other application of ARMA to SHM and damage diagnosis is available in the works of (Entezami and Shariatmadar 2019a; Zheng and Mita 2008; Bao et al. 2013; Carden and Brownjohn 2008; Lakshmi et al. 2016; Entezami et al. 2020d), who attempted to measure the distance of ARMA models fitted to the vibration responses of undamaged and damaged structures. Furthermore, Mosavi et al. (2012) conducted a research work on damage localization under ambient vibration using a Vector AutoRegressive Moving Average (VARMA) model. In their article, the DSFs were obtained from the computation of Mahalanobis distance between the coefficients of VARMA. Recently, Entezami et al. (2020a) exploited the ARMA model and its AR coefficients for feature extraction under Big Data and a large-scale bridge. They demonstrated that this model and coefficient-based feature extraction algorithm is suitable and computationally efficient for using in SHM when large volumes of vibration measurements under ambient excitations are available.

Although the ARMA model is more suitable than the AR representation for feature extraction under ambient vibration, the presence of noise in vibration measurements seriously affects the performance of the ARMA model. To put it another way, when the vibration responses are contaminated by noise and the excitation sources are measurable, it is preferable to use the AR model rather than the ARMA representation (Yao and Pakzad 2012). To deal with the limitation of the ARMA model in noisy conditions, Bao et al. (2013) utilized the auto-correlation function of the normalized acceleration signal as the input data to ARMA rather than using the noisy vibration data.

In connection with the use of the ARX model in the process of feature extraction, Bernal et al. (2012) researched the effect of ARX residuals on damage detection. Kopsaftopoulos and Fassois (2015) presented a residual-based feature extraction by using the ARX model based on the theory of sequential probability ratio test. The main limitation of ARX is the lack of application of this model to the output-only problems. In order to overcome this limitation, Gul and Necati Catbas (2011) proposed a novel methodology based on sensor clustering under the assumption of choosing some vibration responses in a cluster and the remaining responses in the other clusters as input and output data, respectively. In another study for dealing with this obstacle, Roy et al. (2015) presented a feature extraction approach using the ARX model, in which one of the output responses is assumed as the input and the rest is considered as the output.

The ARARX model is a time-invariant linear representation that not only conforms to an AR process but also well suits the output-only SHM condition. The applications of this model to feature extraction can be found in (Sohn and Farrar 2001; Zhang

2007; Farahani and Penumadu 2016; Entezami et al. 2019b, 2020e). Although the AR, ARX, ARMA, and ARARX models are popular among researchers, the use of ARMAX is relatively new. Ay and Wang (2014) applied this model to identify damage by quantifying the acquired set of vibration signals. They presented an iterative algorithm to find optimal orders of ARMAX and reduced the number of orders. Mei et al. (2016) exploited both the coefficients and residuals of the ARMAX model as the DSFs. They indicated that the combination of the ARMAX coefficients and residuals improves the detectability of damage in shear building structures.

Due to the importance of order selection in time series modeling, several authors conducted studies on this significant issue. Figueiredo et al. (2011) assessed the influence of different AR orders on damage detection by using four information criterion techniques. They concluded that the choice of a low order causes obtaining insensitive features and weak damage detectability. Stull et al. (2012) employed information-gap decision theory and robustness curves based on the area under the receiver operating characteristic (ROC) curve versus uncertainty to investigate the effect of AR orders and choose a robust order, which guarantees the extraction of the uncorrelated residuals. Their results demonstrated that the selection a low-order AR model yields the limited number of coefficients and causes the limitation of completely capturing the salient characteristics of the vibration time-domain responses and a poor prediction performance for the other data from the same system. They also concluded that although the use of a high order may lead to good modeling, it might cause an overfitting problem. This problem occurs when redundant and large orders are allocated to a time series model leading to a high-order model with complexity and time-consuming steps of parameter estimation and prediction (Box et al. 2015). In another study, on the order selection of the ARX model by Saito and Beck (2010), the authors concluded that the selection of adequate orders for time-series models is still an important issue in time series modeling. Therefore, this book attempts to propose new and effective order determination techniques in time series analysis to extract reliable features to structural damage from stationary vibration time-domain responses of structures.

The proceeding parts of this chapter of the book are as follows: Sect. 2.2 describes the type and nature of time series data. In Sect. 2.3, various time series models suitable for linear and stationary time series data are discussed and formulated. Sections 2.4 and 2.5 allocate the process of model identification in time series analysis via engineering and statistical aspects. Due to the importance of model identification through the statistical aspect, a conventional graphical approach and an automatic analytical method proposed in this chapter are explained in Sect. 2.5. The proposed algorithms for order determination are presented and explained in Sect. 2.6. The concepts of parameter estimation and different types of feature extraction through time series modeling are explained in Sects. 2.7 and 2.8, respectively. Section 2.9 presents the proposed algorithms of the residual-based feature extraction for SHM. Section 2.10 presents the proposed spectral-based methodology for feature extraction. Eventually, the conclusions of this chapter are summarized in Sect. 2.11.

2.2 Types of Time Series Data

The type and nature of time series data play prominent roles in time series analysis. Before any attempt, it is crucial to understand the nature of the time series. Neglecting this issue, one can encounter serious obstacles to identifying the most appropriate time series model, which should be compatible with data and extract reliable DSFs. On this basis, it is possible to decompose time series data into four groups (Kitagawa 2010):

- Stationary versus non-stationary time series.
- Linear versus nonlinear time series.
- Univariate versus multivariate time series.
- Gaussian versus non-Gaussian time series.

2.2.1 Stationary Versus Non-stationary Time Series

In some cases, time series data contains regularly successive and time-invariant measurements. On this basis, the statistical properties of the data such as the mean, standard deviation, variance, autocorrelation, etc. and their stochastic structures do not change over time. This type of time series is called stationary or time-invariant (Box et al. 2015). On the contrary, non-stationary time series data consist of time-variant measurements so that their statistical properties and stochastic structure change over time. Figure 2.1 illustrates some stationary and non-stationary time series data. In Fig. 2.1a, the samples of data behave in a stationary way, whereas Fig. 2.1b–d indicate the different types of non-stationarity. More precisely, Fig. 2.1b refers to the non-stationarity in mean, while Fig. 2.1d illustrates the non-stationarity in variance. Furthermore, Fig. 2.1c also shows a non-stationary behavior in time series data.

2.2.2 Linear Versus Non-linear Time Series

Depending on the linearity or non-linearity and presence of the input data, a linear time series set is expressible as the output of a linear model, while the output of a non-linear representation is called the nonlinear time series. The graphical analysis of time series highly helps to recognize the linearity or non-linearity of data. For instance, Fig. 2.1a, b indicates that the time series datasets are linear, whereas the time series in Fig. 2.1c is non-linear.

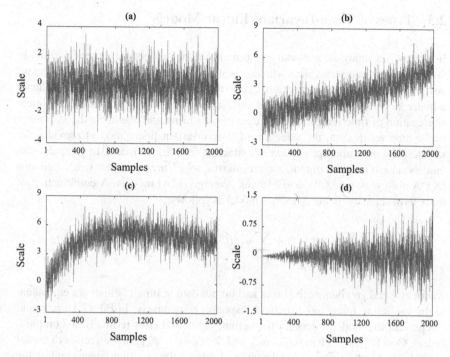

Fig. 2.1 Stationarity and non-stationarity of time series data: **a** stationary time series, **b–d** non-stationary time series

2.2.3 Univariate Versus Multivariate Time Series

Time series data that include a single observation or measurement are called univariate, while multivariate time series data consist of two or more measurements. In a mathematical sense, the univariate and multivariate time series data are expressed by vector and matrix, respectively.

2.2.4 Gaussian Versus Non-Gaussian Time Series

Gaussian time series is one that has a normal distribution, which is the most important and widely used type of probability distribution. Under the central limit theorem, if the samples of time series data are relatively large, this kind of probability distribution is symmetric around the mean (average) of the time series samples (Ross 2014). This signifies that data near the mean are more frequent in occurrence than data far from the mean. In contrast, if the samples of time series data do not conform to the normal distribution, those are non-Gaussian time series. An efficient approach to recognizing the type of distribution of the time series data is kernel density estimation (Scott 2015).

2.3 Types of Time-Invariant Linear Models

In time series analysis, stationary or time-invariant models are generally fitted to the stationary time series data whose statistical properties such as mean, standard deviation, variance, autocorrelation, etc. do not change over time. On the other hand, a linear stochastic model makes the most appropriate model structure for a linear aggregation of time samples. In the case of measuring and acquiring the stationary linear time series data, the selection of time-invariant linear models provides the simplest and most appropriate way of extracting features from data. In general, these models consist of input, output, and error terms, which are equivalent to eXogeneous (X), AutoRegressive (AR), and Moving Average (MA) models. A combination of these terms constructs the ARMAX model as follows:

$$y(t) = \sum_{i=1}^{p} \theta_i y(t-i) + \sum_{j=1}^{r} \varphi_j x(t-j) + \sum_{k=1}^{q} \psi_k e(t-k) + e(t), \qquad (2.1)$$

where $x(t)$ and $y(t)$ denote the input and output data at time t, which are equivalent to the excitation force and structural response, respectively. In Eq. (2.1), p, r, and q represent the model orders for the output, input, and error terms. Based on these orders, $\Theta = [\theta_1 ... \theta_p]$, $\Phi = [\varphi_1 ... \varphi_r]$, and $\Psi = [\psi_1 ... \psi_q]$ are the vectors of model coefficients. Furthermore, $e(t)$ is the residual value at time t, which corresponds to the difference between the measured vibration time series and the predicted one produced by the model. The other polynomial model classes can be established by ignoring each of the input and error terms from the ARMAX representation. In this regard, the ARX model is obtained by eliminating the error term ($q = 0$) from Eq. (2.1) and formulated as follows:

$$y(t) = \sum_{i=1}^{p} \theta_i y(t-i) + \sum_{j=1}^{r} \varphi_j x(t-j) + e(t). \qquad (2.2)$$

In order to construct the ARMA model, it is only necessary to use the output and error polynomials ($r = 0$) leading to the following equation:

$$y(t) = \sum_{i=1}^{p} \theta_i y(t-i) + \sum_{k=1}^{q} \psi_k e(t-k) + e(t). \qquad (2.3)$$

The AR model is taken into account as the simplest polynomial representation, which is formulated by the only output term ($r = q = 0$) as follows:

$$y(t) = \sum_{i=1}^{p} \theta_i y(t-i) + e(t). \qquad (2.4)$$

Apart from such linear polynomial representations, one can develop a two-stage model class named ARARX, which is a combination of AR and ARX. Although the ARARX model is similar to the ARMA representation in the aspect of allocating a polynomial to the error term, it does not use the MA polynomial. In other words, the only AR term is considered to make an ARARX model. Therefore, one can state that the ARARX entirely conforms to the AR process (Ljung 1999). To build the ARARX model, an AR representation is initially fitted to the vibration time-domain response and the model residuals are applied to the ARX model as the input data. Accordingly, the second part of the ARARX model is expressed as:

$$
y(t) = \sum_{i=1}^{\bar{p}} \bar{\theta}_i y(t-i) + \sum_{j=1}^{\bar{r}} \bar{\varphi}_j e(t-j) + \bar{e}(t), \tag{2.5}
$$

where $\bar{p}$ and $\bar{r}$ denote the orders of output and input terms of the ARX representation of the ARARX model; $\bar{\Theta} = [\bar{\theta}_1 \ldots \bar{\theta}_{\bar{p}}]$ and $\bar{\Phi} = [\bar{\varphi}_1 \ldots \bar{\varphi}_{\bar{r}}]$ are the model coefficients. Additionally, $\bar{e}(t)$ represents the residual sequence of the model at time t.

2.4   Model Identification Based on Engineering Aspect

The engineering aspect of the model identification signifies the process of choosing an appropriate class of the time series models without considering the statistical properties of data. In other words, the model identification based on this aspect depends on the types of data acquisition systems. In dynamic tests, there are two general approaches to acquiring vibration data including:

- Input–output or excitation-response data acquisition.
- Output-only or response-only data acquisition.

In the first approach, both the input and output data are measurable and available. In the case of using hammers or shakers for exciting the structures (de Silva 2007), it is feasible to measure both the excitation and response data. This condition refers to the input–output data acquisition system. When the input data is immeasurable and unknown, it is only possible to utilize the output or the response of the structure. Most of the large-scale civil structures are subjected to ambient excitation sources such as wind, traffic loads, human activity, etc. Due to the lack of capability to measure the excitation data, the process of data acquisition is an output-only or response-only condition.

An overview of the time-invariant linear models reveals that the ARX and ARMAX models are normally suitable for the input–output data acquisition system since both of them require the input data for time series modeling. In contrast, one can utilize AR, ARMA, and ARARX for the output-only data acquisition system. An interesting note is that the unmeasurable ambient excitations affect the error term in such a way that a change in the excitation amplitude leads to an alteration in the coefficients of the error polynomial (Carden and Brownjohn 2008). Under such circumstances, therefore, it is necessary to utilize a time series representation that allocates an equation to the error term such as ARMA and ARARX.

2.5   Model Identification Based on Statistical Aspect

The identification of an appropriate model plays an important role in time series analysis. In most cases, the process of model identification is carried out by using the statistical properties of time series samples. This is because there is a broad range of time series representations that are seemingly applicable to the modeling process based on collected data; however, it may not be suitable for the feature extraction resulting from some reasons such as the complexity of modeling, the overfitting problem, and the lack of enough applicability to engineering or vibration-based problems. From a statistical viewpoint, the identified time series model should be compatible with the nature of data (Ljung 1999). Therefore, the preliminary step of the model identification is to realize the nature and behavior of time series with the aid of statistical hypothesis tests. On this basis, it is possible to ignore some irrelevant models without any identification process due to their incompatibility with data. Subsequently, one can use one of the graphical and/or numerical approaches to choosing a time series representation.

2.5.1   A Conventional Approach to Output-Only Conditions

A common graphical tool for the model identification is the Box-Jenkins methodology, which is suitable for choosing a polynomial model among AR, MA, and ARMA via Auto-Correlation Function (ACF) and Partial Auto-Correlation Function (PACF) (Box et al. 2015). In statistics and time series analysis, the ACF measures the correlation between observations of time series data that are separated by h time lags. The formulation of ACF for lag h is:

$$\rho_h = \frac{\vartheta_h}{\vartheta_0}, \tag{2.6}$$

where

$$\vartheta_h = \frac{1}{n-1} \sum_{t=1}^{n-l} (y(t) - \overline{y})(y(t+h) + \overline{y}). \tag{2.7}$$

Furthermore, n is the dimension of time series data. In Eqs. (2.6) and (2.7), ϑ_0 and $\bar{y}$ denote the sample variance and mean of the time series. The PACF gives the partial correlation of time series data with its own lagged values, controlling for the values of the time series at all shorter lags. It contrasts with the ACF, which does not control for other lags. Given the time series $y(t)$, the PACF in the lag h is the correlation between $y(t)$ and $y(t + h)$ with the linear dependence of $y(t)$ and $y(t + 1)$, which is not accounted for by lags 1 to $h - 1$ (Box et al. 2015).

Based on the definitions of the correlation functions and drawing their graphs, one can simply select the most appropriate time-invariant linear model. Regardless of the type of data acquisition system, if the ACF tails off as exponential decay or damped sine wave and the PACF becomes zero after a lag, this implies that the time series data conforms to AR and ARARX. On the contrary, if the PACF tails off as exponential decay or damped sine wave, and the ACF cuts off after a lag, one can select the MA representation. Eventually, if both the ACF and PACF tail off as exponential decay or damped sine waves, ARMA is chosen as the most proper time series model for the input–output data acquisition system (Box et al. 2015).

Despite the simplicity of the Box-Jenkins methodology, it may be difficult and time-consuming to select a proper model for high-dimensional time series data and large datasets. On the other hand, this approach relies strongly on the full inspection of the time series data and the graphs of ACF and PACF. Therefore, the decision about what kind of time-invariant linear representations is suitable for the time series data depends on user inference and expertise (Gómez and Maravall 2000).

2.5.2 An Automatic Approach to Output-Only Conditions

As an alternative way, one can exploit numerical methods that are mainly intended to choose an appropriate time series representation by using statistical criteria. The main advantage of such methods is to create an automatic model identification strategy. The method presented in this section is equivalent to the Box-Jenkins methodology; however, the selection of one of the AR, MA, and ARMA representations is carried out by finite sample criteria based on the ARMA selection (ARMAsel) algorithm (Broersen 2002).

In this algorithm, the automatic model identification approach is based on estimating a large number of AR, MA, and ARMA models, selecting the single best representation for each of the model types, and identifying a model class among the best AR, MA, and ARMA representations. Given n-dimensional time series data, the ARMAsel algorithm estimates AR(p) models with $p = 1,2,...,n/2$ and selects a single best AR representation by Combined Information Criterion (CIC), which is expressed as follows:

$$\text{CIC} = \ln\left(\sigma_e^2\right) + \max\left\{\prod_{i=0}^{p} \frac{1 + \frac{1}{n+1-i}}{1 - \frac{1}{n+1-i}} - 1, \; 3\sum_{i=0}^{p} \frac{1}{n+1-i}\right\}, \quad (2.8)$$

where p denotes the order of AR representation and σ_e^2 is the variance of the model residuals. The best AR model is one that has the smallest CIC value. It is worth remarking that the CIC utilizes a compromise between the finite sample estimator for the Kullback–Leibler information and the optimal asymptotic penalty function (Broersen 2002). In Eq. (2.8), the penalty factor 3 is utilized to reduce the probability of underfitting and overfitting. For the estimation of MA(q) with $q = 1, 2, ..., n/5$, where q represents the order of MA, Generalized Information Criterion (GIC) is used in the following form:

$$\text{GIC} = \ln\left(\sigma_e^2\right) + \frac{3q}{n}. \tag{2.9}$$

Likewise, the same penalty factor as the CIC is applied to the GIC function. Furthermore, it is possible to estimate ARMA($p^*, p^* - 1$) for $p^* = 2, 3, ..., n/10$ by using the GIC equation as follows:

$$\text{GIC} = \ln\left(\sigma_e^2\right) + \frac{3(2p^* - 1)}{n}. \tag{2.10}$$

For both the MA and ARMA representations, the best selection makes the minimum GIC value. Taking the best AR(p), MA(q), and ARMA($p^*, p^* - 1$) models into consideration, the prediction error (PE) of these three representations is computed. Among them, the model class with the smallest PE value is chosen automatically. For MA and ARMA, the equation of PE is given by:

$$PE(a) = \sigma_e^2 \left(\frac{1 + \frac{a}{n}}{1 - \frac{a}{n}} \right), \tag{2.11}$$

where a is the total number of estimated parameters (coefficients) in those models. For the AR(p) representations, the PE equation is expressed as:

$$PE(p) = \sigma_e^2 \left(\prod_{i=0}^{p} \frac{1 + \frac{1}{n+1-i}}{1 - \frac{1}{n+1-i}} \right). \tag{2.12}$$

It is important to note that if the number of the estimated coefficients (the orders) of the AR model are smaller than $n/10$, the PE values of the ARMA and AR models respectively, presented in the Eqs. (2.11) and (2.12), are the same amounts. Furthermore, in order to reduce the computational time, it would be very appropriate to restrict the orders of AR, MA, and ARMA models to 500 or 1000 numbers rather than using the amounts of $n/2$, $n/5$, and $n/10$ (Broersen 2002). Note that the number n can be any value.

2.6 Proposed Order Selection Algorithms

The process of order determination is a crucial step in time series modeling because an inadequate order can result in an underfitting problem and an inappropriate model with poor performance or goodness-of-fit (Box et al. 2015; Fassois and Sakellariou 2009). In other words, if the orders are not sufficient, it is possible to fit a low-order model, in which case the underfitting problem occurs. Depending on the type of time series representation, the model orders specify how many unknown parameters or coefficients should be allocated to predict the response of the structure and model the response of interest. On this basis, the choice of adequate orders depends strongly on the residuals of the time series model (Hyndman and Athanasopoulos 2014). From a statistical point of view, the orders should enable the time series models to produce uncorrelated (independent) residuals as white noise (Bisgaard and Kulahci 2011; Box et al. 2015). In other words, the uncorrelatedness of residual samples is a significant criterion for understating the model accuracy and adequacy. Any time series model that does not satisfy this requirement is not accurate and adequate and should be modified (Box et al. 2015). Throughout this research work, an adequate and accurate time series model is one that generates uncorrelated residuals. This means that the orders used in that model are sufficient (i.e. neither very small nor very large) without underfitting and overfitting problems.

In time series analysis, information criterion techniques such as Akaike Information criterion (AIC), Corrected Akaike Information Criterion (AICC), and Bayesian Information Criterion (BIC) are usually applied to determine the orders of time series models (Box et al. 2015). Given an n-dimensional vibration response and a model with a coefficient number (e.g. $a = p + r + q$ for ARMAX, $a = p + q$ for ARMA, $a = p + r$ for ARX, and $a = p$ for AR), the AIC is given by:

$$AIC = -2\ln(L_{max}) + 2a, \tag{2.13}$$

where L_{max} represents the maximum value of the likelihood estimate of the model, which is the value that maximizes L for the given data (Brockwell and Davis 2016). A drawback of AIC is a tendency to overestimate a, which leads to an overfitting problem (Bisgaard and Kulahci 2011). To remedy this limitation, the AICC and BIC enhance the AIC by adding rigorous penalty terms in the following forms:

$$AICC = -2\ln(L_{max}) + 2a\left(\frac{n}{n-a-1}\right), \tag{2.14}$$

$$BIC = -2\ln(L_{max}) + a\ln(n). \tag{2.15}$$

The main properties of AICC and BIC lie in the fact that the AICC is usually suitable for smaller samples, while the BIC is well suited for larger sequences (Bisgaard and Kulahci 2011). The process of order selection by the information criterion techniques is based upon examining a wide range of orders for different model classes and computing their AIC, AICC, and BIC values. An adequate model order without the underfitting and overfitting problems is one that has the smallest amount of these criteria. It should be mentioned that the model order selection is implemented by one of the AIC, AICC, and BIC approaches. Due to some advantages of BIC (i.e. the lack of occurrence of overfitting problem and being suitable for large datasets), it is preferable to utilize this technique to determine the model orders.

2.6.1 Robust Order Selection by an Iterative Algorithm

This section presents the main concepts of robust order selection. The robustness of model order means that the order of interest enables the time series model to generate uncorrelated residuals. The main idea behind the proposed iterative method is to examine the model residuals by Ljung-Box Q-test (LBQ). This is a portmanteau test, which evaluates whether the model residuals are uncorrelated based on the null or alternative hypotheses. The Ljung-Box statistic or statistical quantity is given by:

$$Q_{LB} = n(n+2) \sum_{h=1}^{L} \frac{\rho_h^2}{n-h}. \tag{2.16}$$

Similar to all statistical hypothesis tests, the LBQ test gives some numerical amounts for making a decision. These amounts include a probability value (p-value), a critical value (c-value), and the test statistic (Q_{LB}) under a significance level (α). Accordingly, the test decision can be classified as the null ($\mathbb{H}_0$) and alternative ($\mathbb{H}_1$) hypotheses. The null hypothesis means that the test accepts the underlying premise or theory of interest (e.g. the uncorrelatedness of the residual samples in the LBQ test); otherwise, the test rejects it and the decision becomes the alternative hypothesis. For the LBQ test, the null hypothesis occurs when the test statistic is less than the c-value and/or the p-value is larger than the significance level. Such conditions imply that the model residuals are uncorrelated. On the contrary, if the Q_{LB} is greater than the c-value and/or the p-value is smaller than the significance level, the LBQ test rejects the null hypothesis implying that the model residuals are still correlated. With these descriptions, one can argue that the p-value, c-value, and Q_{LB} in the LBQ test are suitable criteria for checking the correlation of the model residuals.

It should be mentioned that the c-value is related to the significance level and the degrees of freedom for the χ^2-distribution of the test statistic. In statistics, the degree of freedom is the number of values in the final calculation of a statistic that are free to vary. In the LBQ test, the degrees of freedom are identical to the number of lagged terms for consisting of the calculation of the test statistic. A common way to select the degrees of freedom is to calculate $min(20, n-1)$ (Box et al. 2015). Hence, the c-value is determined by considering the significance level and degrees of freedom based on the χ^2 inverse cumulative distribution function. For the 5%

significance level ($\alpha = 0.05$) and 20 degrees of freedom, the c-value is identical to 31.4104 (Brockwell and Davis 2016).

The process of iterative order selection begins with choosing sequential model orders (p_i, r_j, and q_k, where i, j, k = 1,2,...) from one to a number that the model residuals become uncorrelated. Fitting a stochastic time-invariant model to the measured vibration data, the model coefficients are estimated in each iteration. After extracting the model residuals, the LBQ test is applied to analyze their correlation.

If the residual sequences of the model of interest are uncorrelated ($Q_{LB} \leq c$-value and/or p-value > α), the iteration number (p_i, r_j, and q_k) at the end of the iterative algorithm are chosen as the robust orders; otherwise ($Q_{LB} > c$-value and/or p-value $\leq \alpha$), the iterative algorithm continues by adopting a new iteration number. Note that the selection of model orders depends on the result of model identification. For example, if the AR model is chosen as the most appropriate time series representation, it is only necessary to implement the iterative algorithm of order determination for p_i. Figure 2.2 illustrates the flowchart of the proposed method of iterative order selection. The main advantage of this approach is that it gives robust orders in the sense that they always guarantee the uncorrelatedness of the residual samples and the model accuracy and adequacy.

Fig. 2.2 The flowchart of the proposed iterative algorithm for the robust order selection

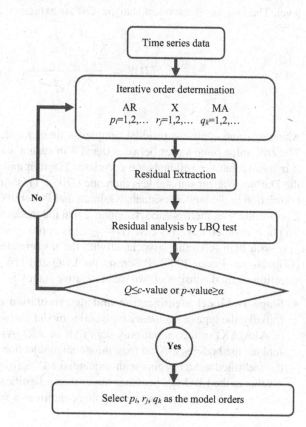

2.6.2 Robust and Optimal Order Selection by a Two-Stage Iterative Algorithm

There is no doubt that the application of the LBQ test to the iterative algorithm of the previous method provides a great opportunity to select a robust order, which enables the time-series models to extract uncorrelated residuals. The main limitation of this approach is to choose high orders that increase the probability of the overfitting problem. To avoid encountering this issue, this section proposes a two-stage iterative algorithm to select not only a robust order but also an optimal one such that the model is able to produce uncorrelated residuals with a few orders. The fundamental idea behind this method relies upon using the residual analysis by the LBQ test in the first iterative stage and the Durbin-Watson hypothesis test in the second iterative process. In fact, the use of the second iterative algorithm leads to providing smaller orders than the previous iterative method.

Similar to the LBQ test, which has been discussed in the previous section, the Durbin–Watson hypothesis test yields a statistic, which can be used to assess whether the residual samples are correlated or uncorrelated (Durbin and Watson 1950). Similarly, this test gives a p-value based on a specific confidence level and a significance level. The Durbin–Watson test statistic, DW, is expressed as:

$$DW = \frac{\sum_{t=1}^{n-1} (e_{t+1} - e_t)^2}{\sum_{i=1}^{n} e_t^2}, \tag{2.17}$$

where e_{t+1} and e_t are the residual samples at times t and $t+1$ for $t = 1, 2, ..., n$. The DW value often varies between 0 and 4 in such a way that DW < 2 or p-value $< \alpha$ means that the residuals are correlated (Durbin and Watson 1950). Note that the Durbin–Watson statistic less than one ($DW < 1$) is indicative of the substantial correlation in the model residuals (Gujarati and Porter 2003). By contrast, if $DW \geq$ 2 or p-value $\geq \alpha$, there is enough evidence that the model residuals are uncorrelated. To put it another way, this case ($DW \geq 2$) is approximately equivalent to p-value $\geq \alpha$ with high reliability associated with the uncorrelatedness of model residuals (Gujarati and Porter 2003). Based on the LBQ and DW tests, the proposed order selection method consists of two main iterative stages as follows:

- **Stage 1- Model identification and determination of the maximum order**: Initially, the type of stationary time-series model for the input–output data (ARX or ARMAX) or the output-only data (AR or ARMA) is identified by the Box-Jenkins methodology. After that, the iterative selection of the maximum order of the identified model begins with sequential orders from one to a number that its p-value in the LBQ-test becomes larger than a significance level. Hence, the iteration number terminated at the stopping conditions p-value $\geq \alpha$ or $Q_{LB} \leq c$-value

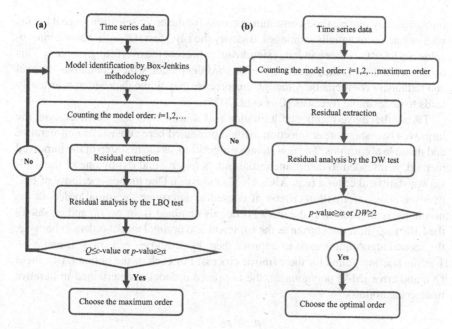

Fig. 2.3 The flowcharts of the proposed two-stage iterative algorithm of model order selection: **a** determining the maximum order, **b** choosing the optimal order

refers to the maximum order required by the time-series model for extracting the uncorrelated residuals.

- **Stage 2- Selection of the optimal order**: As stated earlier, in order to prevent the overfitting problem caused by selecting the high orders via the LBQ test, one attempts to choose robust and optimal orders with the focus on extracting the uncorrelated residuals and providing fewer orders. On this basis, the number of orders begins with one to the maximum order gained by the preceding stage. The stopping conditions in the current stage are the outputs of the Durbin-Watson test including p-value $\geq \alpha$ or $DW \geq 2$. Similarly, with the previous step, the iteration number that satisfies the stopping conditions is chosen as the optimal order; otherwise, a new order should be allocated to the time series model. Figure 2.3 illustrates the flowchart of the two-stage iterative algorithm for choosing a robust and optimal order.

2.6.3 Robust and Optimal Order Selection by an Improved Two-Stage Algorithm

Although the proposed iterative algorithm for the determination of robust orders presented in Sect. 2.6.1 is innovative and efficient, it may suffer from some obstacles to select optimal orders for high-dimensional time series datasets (Big data). This limitation may make a time-consuming process for the order selection since it begins with one and sequentially continues to satisfy the LBQ test ($Q_{LB} \leq c$-value and/or p-value $> \alpha$). On the other hand, it is significant to prevent the overfitting problem. This occurs when the obtained orders make a high-order model, which contains additional and redundant coefficients. Although this process may fit the data better, it seriously leads to undesirable forecasts (Box et al. 2015).

Taking the above-mentioned limitation and shortcoming into consideration, the improved two-stage order selection method presented here consists of non-iterative and iterative algorithms. The non-iterative algorithm, as the first step of the improved method, is intended to determine initial orders (p_0, r_0, and q_0) by one of the well-known statistical criteria (e.g. AIC, AICC, and BIC). Due to the superiority of BIC over the other information criteria, it is used to determine the initial orders in the non-iterative algorithm. If the model residuals obtained from p_0, r_0, and q_0 satisfy the LBQ test, these are chosen as the sufficient and optimal model orders, otherwise, the second algorithm is used to improve them by setting $p_0 + 1$, $r_0 + 1$, and $q_0 + 1$ as the starting points for the iterative process. For each of the output (AR), input (X), and error (MA) polynomials, the improved orders are determined in iterative manners as follows:

$$
\begin{aligned}
p_i &= p_0 + i, \\
r_j &= r_0 + j, \\
q_k &= q_0 + k,
\end{aligned}
\tag{2.18}
$$

where $i, j, k = 1,2,\ldots$ The iterative process continues to fulfill the LBQ test. Afterward, the overfitting problem is evaluated by R-squared (R^2) and adjusted R-squared (Adj-R^2) statistics (Montgomery et al. 2015) in the second algorithm. Assuming that $\mathbf{y}(t) = [y(1)\, y(2) \ldots y(n)]$ is a response vector and $\mathbf{e}(t) = [e(1)\, e(2) \ldots e(n)]$ refers to the residual vector of a time series model with a coefficient number, the R-squared statistic is given by:

$$
R^2 = 1 - \frac{\sum\limits_{t=1}^{n} e(t)^2}{\sum\limits_{t=1}^{n} (y(t) - \overline{y})^2},
\tag{2.19}
$$

where $\overline{y}$ denotes the mean of $\mathbf{y}(t)$. The R^2 always varies from zero to one in such a way that a large value close to one suggests a good fit to the time series data. However, it does not necessarily imply that the time series model is appropriate. This is because an additional order applied to the model will never lead to a reduction in R^2, even in conditions that the order of interest is not statistically significant. Under such circumstances, the occurrence of the overfitting problem is common. To circumvent this drawback, Adj-R^2 is defined as follows:

$$Adj-R^2 = 1 - \frac{\left(\frac{1}{n-a}\right) \sum\limits_{t=1}^{n} e(t)^2}{\left(\frac{1}{n-1}\right) \sum\limits_{t=1}^{n} (y(t) - \overline{y})^2}. \tag{2.20}$$

Similar to R^2, the adjusted R-squared statistic varies in the range of zero to one; however, it may give negative values. In general, Adj-R^2 will not always increase by adding an additional order. If unnecessary orders are added to the model, the value of the adjusted R-squared statistic will often decrease. As a result, a large amount of Adj-R^2 close to one is indicative of a good fit. For avoiding and checking the overfitting problem, if the value of the adjusted R^2 statistic is positive and not much less than the R^2 statistic, it can be argued that the overfitting problem does not occur (Montgomery et al. 2015). As a result, one can select p_i, r_j, and q_k as sufficient and optimal model orders. Note that if the R-squared and adjusted R-squared statistics have serious differences or Adj-R^2 becomes negative, it is preferable to alter the model class and select a more parsimonious time series representation or an accurate and adequate model with fewer orders. The flowchart of the improved iterative method for selecting the optimal and robust orders is depicted in Fig. 2.4. The great merit of this method is to provide optimal and robust orders, which means that these not only guarantee the model adequacy and accuracy but also avoid the overfitting problem.

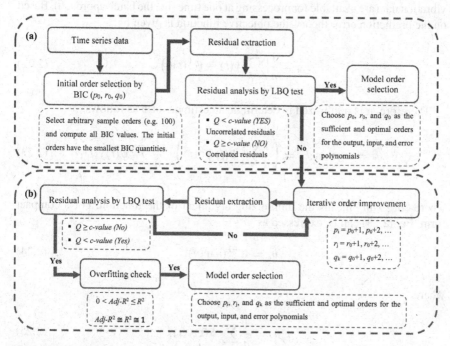

Fig. 2.4 The flowchart of the proposed improved two-stage algorithm of model order selection: **a** the non-iterative algorithm, **b** the iterative algorithm

2.7 Parameter Estimation

The term parameter estimation refers to the process of using sample data to estimate the unknown coefficients of a time series model (Ljung 1999; Box et al. 2015). This process is faced with the problem of how to use the information contained in the time series data to properly estimate the model coefficients. The unknown values of the model coefficient amounts should be able to describe the data and make a good model. Among the well-known parameter estimation techniques including Least Squares (LS), Maximum Likelihood (ML), and Bayesian, the first one is taken into account as a simple and robust estimation approach to engineering applications (Cooper and Worden 2000). The LS technique is typically divided into the non-recursive and recursive algorithms (Young 2011). This section describes the non-recursive LS estimation for the AR model. Given $\psi(t) = [-y(t-1) -y(t-2) \ldots -y(t-p)]^T$ as the regression vector of the AR model, Eq. (2.4) can be rewritten as follows (Ljung 1999):

$$y(t) = \psi^T(t)\theta + e(t), \tag{2.21}$$

where $\psi^T(t)\theta$ is the vector inner product. The main goal of the non-recursive LS technique is to minimize an objective function under the assumption that all the measured vibration data are available for processing at one time (i.e. the batch approach). Based on the prediction error theory, the objective function is given by:

$$\mathcal{F}_n = \frac{1}{n}\sum_{t=1}^{n}\left(y(t) - \psi^T(t)\theta\right)^2. \tag{2.22}$$

Therefore, it is minimized to determine an estimate of θ as follows:

$$\hat{\theta}_n = \arg\min \mathcal{F}_n = \left(\frac{1}{n}\sum_{t=1}^{n}\psi(t)\psi^T(t)\right)^{-1}\frac{1}{n}\sum_{t=1}^{n}\psi(t)y(t), \tag{2.23}$$

where "*arg min*" means the minimizing argument of the function. For a compact form, Eq. (2.23) can be expressed as:

$$\hat{\theta}_n = \mathbf{R}^{-1}(n)\mathbf{r}(n), \tag{2.24}$$

where

$$\mathbf{r}(n) = \frac{1}{n}\sum_{t=1}^{n}\psi(t)y(t), \tag{2.25}$$

$$\mathbf{R}(n) = \frac{1}{n} \sum_{t=1}^{n} \boldsymbol{\psi}(t)\boldsymbol{\psi}^T(t).$$ (2.26)

2.8 Types of Feature Extraction via Time Series Analysis

Feature extraction by time series analysis is typically categorized as coefficient-based or residual-based algorithms (Figueiredo et al. 2009). The coefficient-based feature extraction (CBFE) relies on estimating the model coefficients in the training and inspection phases (e.g. the undamaged and damaged conditions) by using orders that are generally obtained from the undamaged states with different environmental and/or operational variability conditions in the training period. In contrast, the residual-based feature extraction (RBFE) is based on extracting the model residuals in both phases with the aid of the model orders and coefficients gained by the normal conditions. The main idea behind the RBFE algorithm is that the model used in the undamaged state will no longer predict the responses of the damaged structure due to discrepancies in the structural properties and responses caused by damage occurrence. In such a case, the model residuals regarding the damaged structure will increase (Fassois and Sakellariou 2009; Figueiredo et al. 2011; Mei et al. 2016). Although the use of CBFE is common and yields reliable and accurate results, the major benefit of the RBFE algorithm is that the order determination and parameter estimation procedures are not implemented in the inspection phase (Fassois and Sakellariou 2009). This advantage leads to the establishment of an unsupervised learning strategy in the feature extraction. For the sake of clarification, Fig. 2.5 illustrates the flowcharts of the CBFE and RBFE algorithms. A note regarding the CBFE algorithm is that among the coefficients of AR, X, and MA terms (i.e. $\boldsymbol{\Theta}$, $\boldsymbol{\Phi}$, and $\boldsymbol{\Psi}$), the coefficients of the AR term are generally used as the DSFs due to their sensitivity to structural changes. In contrast, the coefficients of input term (X) do not have any influence on the structural properties and depend only on the source of excitation (Yao and Pakzad 2012).

2.9 Proposed RBFE Methods

2.9.1 A Developed RBFE Approach

Taking the importance of model order into account, the conventional RBFE approach is enhanced to provide a more appropriate algorithm for feature extraction. On this basis, a developed residual-based feature extraction approach called DRBFE is proposed here that contains two general stages. The first one is implemented by using the information on the normal or undamaged conditions of the structure. In

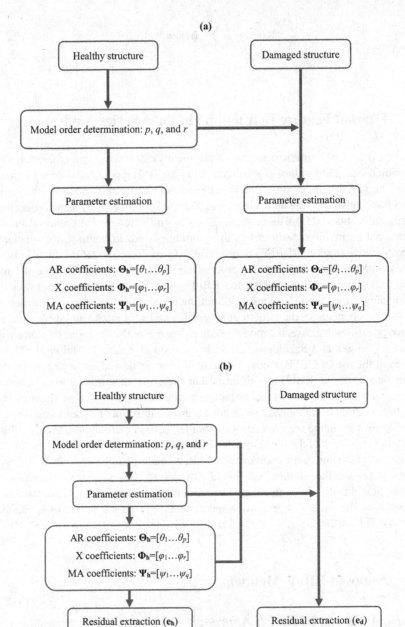

Fig. 2.5 Flowcharts of conventional feature extraction by time series analysis: **a** CBFE, **b** RBFE

this stage, one attempts to identify an appropriate model based on the Box-Jenkins methodology (the conventional model identification approach), determine adequate and accurate orders, estimate the model parameters, and then extract the uncorrelated residuals of the identified model at each sensor as the DSFs. On the contrary, the second stage is concerned with the damaged state of the structure. In this stage, the model characteristics (i.e. the orders and parameters) obtained from the normal condition are applied to extract the residual sequences associated with the damaged state.

- **Step 1—Model identification**: Based on the description in Sect. 2.5.1, the most proper time series representation ($\mathcal{M}$) is identified by using the Box-Jenkins methodology. This model can be one of the time-invariant linear models for output-only conditions such as AR, ARMA, and ARARX.
- **Step 2—Initial order determination**: In this step, it is attempted to determine initial orders (p_0, q_0, and r_0) at each sensor by the BIC technique or Eq. (2.15).
- **Step 3—Improved order determination**: Although the information criteria are usually applied to choose the orders of time series representations, the uncorrelatedness of the residual samples gained by them may not be fully satisfied. Hence, the initial orders are developed to achieve the improved orders (p_i, q_i, and r_i). The development is based on evaluating the correlation of residual sequences by the ACF. If the ACF's values are roughly located between the upper and lower bounds of a confidence interval, one can understand that the model residuals are uncorrelated and p_i, q_i, and r_i are chosen as the improved orders; otherwise, they should be increased.
- **Step 4—Maximum order selection**: For the damage detection problems, it is better to use features (either the model parameters or the model residuals) with the same dimensions for both the undamaged and damaged states. To provide this requirement and deal with the inequality of feature dimensions, the maximum number of the improved orders is selected to utilize in the parameter estimation and residual extraction. In this step, p_m, q_m, and r_m denote the maximum orders for the output, error, and input terms, respectively.
- **Step 5—Modeling by the maximum orders**: The time series representation identified in the first step uses the maximum orders to model the vibration responses of all sensors. The main property (advantage) of this kind of modeling is to guarantee the extraction of uncorrelated residuals from all sensors.
- **Step 6—Parameter estimation**: The unknown model parameters ($\boldsymbol{\Theta}_m$, $\boldsymbol{\Phi}_m$, and $\boldsymbol{\Psi}_m$) are estimated by the non-recursive LS technique.
- **Step 7—Residual extraction for the undamaged state**: The uncorrelated model residuals at each sensor are extracted as the DSFs for the normal condition.
- **Step 8—Modeling by the information of the undamaged state:** In this step, the maximum orders and the model parameters obtained from the normal condition are employed to model the vibration responses of the damaged structure.
- **Step 9—Residual extraction for the damaged state**: Similar to step 7, the model residuals at each sensor are extracted as the DSFs for the damaged condition.

2.9.2 A Fast RBFE Approach

Even though the conventional RBFE algorithm has its advantages, it may be time-consuming under vibration datasets with high levels of dimensionality and large volumes. On this basis, a fast residual-based feature extraction (FRBFE) is proposed to overcome this limitation. This technique is divided into two general stages including training (Stage I) and monitoring (Stage II), each of which consists of the initial and iterative steps. In the first stage, one attempts to choose an optimal and robust order by improving the iterative order selection approach as discussed in Sect. 2.6.1 and then extract the model residuals as the main DSFs in the last iteration. The second stage exploits the model information of the previous stage in the training phase and repeats the iterative process (i.e. without order determination and parameter estimation) to extract the model residuals as the main DSF of the current structural state (damaged structure).

- **Stage I—Model order selection and residual extraction in the training phase:**
  Similar to the original algorithm of selecting the robust order, the improved method lies in analyzing the residual sequences of the model of interest, which can be one of the time-invariant linear representations, by the LBQ test. The main idea behind the improved approach is to use the correlated residual sequences as a new time series data in the iterative process instead of applying the measured vibration response. As important merit, this process leads to a fast order selection with less computational time than the original technique. For the sake of convenience, Fig. 2.6 depicts the schematic representation of the initial and iterative steps of

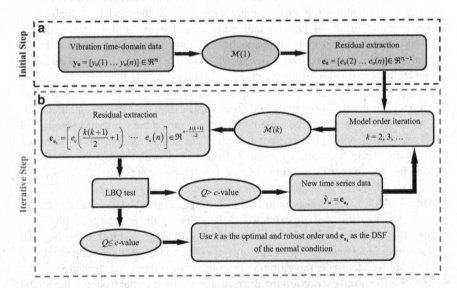

Fig. 2.6 The flowchart of the initial and iterative steps of the proposed FRBFE in the training phase: **a** the initial step, **b** the iterative step

the proposed FRBFE method in the training phase. In the following, these steps are described in detail.

Initial step: Assume that $\mathbf{y_u} = [y_u(1) \ldots y_u(n)] \in \mathfrak{R}^n$ is the vector of measured vibration time-domain response associated with the undamaged or normal condition. In the initial step, a time series model named as $\mathcal{M}$ with order one (i.e. $\mathcal{M}(1)$, which can be AR(1), ARMA(1,1), ARARX(1,1,1)) is fitted to the response vector in order to extract the model residuals as $\mathbf{e_u} = [e_u(2)\ldots e_u(n)] \in \mathfrak{R}^{n-1}$. Note that before extracting the residuals, the model coefficients are estimated by the non-recursive LS technique. Because the residual sequence of the first order, $e_u(1)$, approximately corresponds to zero, one can neglect it. In most cases of high-dimensional vibration data, where n is a relatively large number, the residuals of $\mathcal{M}(1)$ are correlated. Hence, $\mathbf{e_u}$ is taken into account as a new time series dataset to use in the iterative step. The central idea is that if the residuals are correlated, this means that some information is present that is not captured by the model. It is worth mentioning that if the residual analysis via LBQ test on the residuals of $\mathcal{M}(1)$ indicates that those are uncorrelated, which is an almost rare event, one can use the orders obtained from this stage and $\mathbf{e_u}$ as the selected model orders and the main DSF of the normal condition, respectively.

Iterative step: By increasing the model orders ($k = 2,3,\ldots$), $\mathcal{M}(k)$ is fitted to the correlated residual sequences obtained from the previous step (if $k = 2$) or the previous iteration (if $k > 2$) in order to extract the new residual vector $\mathbf{e_{uk}} = [e(k(k + 1)/2 + 1)\ldots e(n)] \in \mathfrak{R}^{n-k(k+1)/2}$ in the kth iteration. Once again, the non-recursive LS technique is applied to estimate the model coefficients in each iteration before the residual extraction. Since the first k samples of the residual vector of $\mathcal{M}(k)$ are zero, similar to the previous step, those are eliminated from $\mathbf{e_{uk}}$ in each iteration. If the sequences of $\mathbf{e_{uk}}$ satisfy the null hypothesis of the LBQ test ($Q \leq c$-value), the iterative process terminates; otherwise ($Q > c$-value), the model orders or iteration number (k) increases by considering the correlated residual vector as a new dataset. It is important to note that $Q \leq c$-value is a stopping condition for terminating the iterations. Accordingly, the iterative process continues until $Q \leq c$-value, in which case the number of iterations refers to the optimal and robust orders. Eventually, the uncorrelated residual vector ($\mathbf{e_{uk}}$) of the last iteration is used as the main DSF of the training phase. Note that the model coefficients of each iteration should be stored to utilize them in the residual extraction of the monitoring phase.

- **Stage II—Residual extraction in the monitoring phase:** In order to obtain the DSF of the current structural state or damaged structure, it is only necessary to repeat the initial and iterative steps of the previous algorithm for extracting the residual sequences of $\mathcal{M}(k)$.

Initial step: It is assumed that $\mathbf{y_d} = [y_d(1) \ldots y_d(n)] \in \mathfrak{R}^n$ is the vector of measured vibration time-domain response for the current structural state. Fitting $\mathcal{M}(1)$ to this data, the residual vector is $\mathbf{e_d} = [e_d(2)\ldots e_d(n)] \in \mathfrak{R}^{n-1}$. In case of using the residual vector $\mathbf{e_u}$ in the initial step of the training phase as the main DSF of the normal condition, one should apply $\mathbf{e_d}$ for the current state; otherwise, it is used as the new time series data in the iterative step.

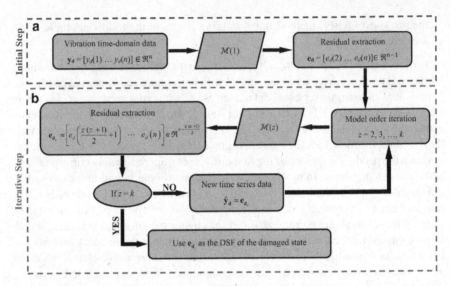

Fig. 2.7 The flowchart of the initial and iterative steps of the proposed FRBFE method in the monitoring phase: **a** the initial step, **b** the iterative step

Iterative step: The model order iteration in this step begins with 2 and ends with k; that is, $z = 2, 3, \ldots, k$. Using the model coefficients estimated from the previous algorithm, the residuals of $\mathcal{M}(z)$ are extracted in the zth iteration. If $z = k$, the residual vector $\mathbf{e}_{d_z} = [e_d(\frac{z(z+1)}{2}+1)\ldots e_d(n)] \in \Re^{n-z(z+1)/2}$ is utilized as the DSF of the current structural condition; otherwise, the iterative process continues by choosing $\mathbf{e}_{d_z}$ as the new time series data $(\hat{\mathbf{y}}_d)$. Figure 2.7 illustrates the flowchart of the initial and iterative steps of the proposed FRBFE method in the monitoring phase.

2.10 Proposed Spectral-Based Method

The spectral analysis provides many approaches that characterize the frequency content of a signal. This analysis is based on estimating the PSD of a signal from its time-domain representation. Spectral density characterizes the frequency content of a signal or a stochastic process (Castanié 2013; Stoica and Moses 1997). The spectral analysis is often carried out by nonparametric and parametric methods. The nonparametric techniques such as FFT-based Welch's method or periodogram do not require prior knowledge and details about the signal data. The main advantage of non-parametric methods is the applicability to any kind of signal. Parametric methods such as Burg's, covariance, and MUSIC are model-based approaches that incorporate prior knowledge of the signal and can yield more accurate spectral estimates. The model for generating the signal can be constructed with several parameters that can be estimated from the observed data. From the model and estimated parameters, the algorithm computes the power density spectrum implied by the model. These methods estimate the PSD by first estimating the parameters of the linear system that hypothetically generates the signal. The superiority of the parametric methods over the non-parametric techniques is concerned with their tendency to produce better results and high resolutions (Candy 2005).

The most commonly used linear system model in the parametric spectral-based approaches is the AR representation, which is known as an all-pole model in signal processing sense. Based on Eq. (2.4), where is formulated an AR model with p orders and $\theta_1 \dots \theta_p$ coefficients, it is possible to estimate the AR spectrum (Yao and Pakzad 2012). On the other hand, one of the effective approaches to estimating the AR spectrum among various parameter estimation techniques is Burg's method. The main advantages of the Burg's method compared to the other AR spectral estimation techniques are resolving closely spaced sinusoids in signals with low noise levels, and estimating short data records, in which case the AR spectral estimates are very close to the true values (Bos et al. 2002). In addition, it ensures a stable AR model and provides a computationally efficient method for parameter estimation. The Burg's is based on minimizing the forward and backward prediction errors while satisfying the Levinson-Durbin recursion (Stoica and Moses 1997). This method avoids calculating the autocorrelation function and instead estimates the reflection coefficients directly. More speaking, the pth reflection coefficient is a measure of the correlation between $y(t)$ and $y(t - p)$ after the correlation due to the intermediate observations $y(t - 1)$ ... $y(t - p + 1)$ has been filtered out. Reflection coefficients can be transformed into autoregressive parameters by applying the Levinson–Durbin recursion formula (Stoica and Moses 1997). Due to the advantages of Burg's method, this book uses this approach to estimate the AR spectrum $P(\omega)$ in the following form:

$$P(\omega) = \frac{\sigma_e^2}{\left| 1 - \sum_{k=1}^{p} \theta_k e^{-j\omega k} \right|^2} \tag{2.27}$$

where σ_e^2 denotes the variance of the AR model residuals.

2.11 Conclusions

This chapter of the book has presented the proposed methods based on time series modeling for feature extraction under the stationarity and linearity of the measured vibration time-domain responses. It has been described the importance and effectiveness of time series analysis and modeling for feature extraction. Some important challenges and limitations of this approach have been mentioned to clarify the reasons for the proposed methods. Among them, the identification of a proper time series representation is the initial step of time series modeling. Determination of the optimal and robust order is taken into account as the crucial issue in time series modeling so that it should enable the model of interest to generate uncorrelated residuals without the overfitting problem. Consideration of the EOV is the other important issue for feature extraction by time series modeling. Aiming at dealing with the above-mentioned issues and some limitations in the conventional approaches, innovative order selection algorithms, and novel feature extraction methods have been proposed in this chapter.

References

Ay AM, Wang Y (2014) Structural damage identification based on self-fitting ARMAX model and multi-sensor data fusion. Struct Health Monit 13(4):445–460

Bao C, Hao H, Li Z-X (2013) Integrated ARMA model method for damage detection of subsea pipeline system. Eng Struct 48:176–192

Bernal D, Zonta D, Pozzi M (2012) ARX residuals in damage detection. Struct Control Health Monit 19(4):535–547

Bisgaard S, Kulahci M (2011) Time series analysis and forecasting by example. Wiley

Bos R, De Waele S, Broersen PM (2002) Autoregressive spectral estimation by application of the Burg algorithm to irregularly sampled data. IEEE Trans Instrum Meas 51(6):1289–1294

Box GEP, Jenkins GM, Reinsel GC, Ljung GM (2015) Time series analysis: forecasting and control, 5th edn. Wiley

Brockwell PJ, Davis RA (2016) Introduction to time series and forecasting. Springer

Broersen PM (2002) Automatic spectral analysis with time series models. IEEE Trans Instrum Meas 51(2):211–216

Candy JV (2005) Model-based signal processing, vol 36. Wiley

Carden EP, Brownjohn JM (2008) ARMA modelled time-series classification for structural health monitoring of civil infrastructure. Mech Syst Sig Process 22(2):295–314

Castanié F (2013) Spectral analysis: parametric and non-parametric digital methods. Wiley

Cooper JE, Worden K (2000) On-line physical parameter estimation with adaptive forgetting factors. Mech Syst Sig Process 14(5):705–730

Datteo A, Lucà F (2017) Statistical pattern recognition approach for long-time monitoring of the G. Meazza stadium by means of AR models and PCA. Eng Struct 153:317–333

de Silva CW (2007) Vibration monitoring, testing, and instrumentation. CRC Press

de Lautour OR, Omenzetter P (2010) Damage classification and estimation in experimental structures using time series analysis and pattern recognition. Mech Syst Sig Process 24(5):1556–1569

Durbin J, Watson GS (1950) Testing for serial correlation in least squares regression I. Biometrika 37:409–428

Dyke SJ, Bernal D, Beck J, Ventura C (2003) Experimental phase II of the structural health monitoring benchmark problem. In: Proceedings of the 16th ASCE engineering mechanics conference

Entezami A, Shariatmadar H (2017) An unsupervised learning approach by novel damage indices in structural health monitoring for damage localization and quantification. Struct Health Monit 17(2):325–345

Entezami A, Shariatmadar H (2019) Damage localization under ambient excitations and non-stationary vibration signals by a new hybrid algorithm for feature extraction and multivariate distance correlation methods. Struct Health Monit 18(2):347–375

Entezami A, Shariatmadar H (2019) Structural health monitoring by a new hybrid feature extraction and dynamic time warping methods under ambient vibration and non-stationary signals. Measurement 134:548–568

Entezami A, Shariatmadar H, Karamodin A (2018a) Data-driven damage diagnosis under environmental and operational variability by novel statistical pattern recognition methods. Struct Health Monit 18(5–6):1416–1443

Entezami A, Shariatmadar H, Karamodin A (2018b) An improvement on feature extraction via time series modeling for structural health monitoring based on unsupervised learning methods. Scientia Iranica

Entezami A, Sarmadi H, Salar M, Behkamal A, Arslan AN, De Michele C (2019a) A novel structural feature extraction method via time series modelling and machine learning techniques for early damage detection in civil and architectural buildings. In: International conference on emerging technologies in architectural design (ICETAD2019), pp 72–78

Entezami A, Shariatmadar H, Mariani S (2019b) A novelty detection method for large-scale structures under varying environmental conditions. In: Sixteenth international conference on civil, structural & environmental engineering computing (CIVIL-COMP2019), 16–19 September, 2019, Riva del Garda, Italy

Entezami A, Sarmadi H, Behkamal B, Mariani S (2020) Big data analytics and structural health monitoring: a statistical pattern recognition-based approach. Sensors 20(8):2328

Entezami A, Shariatmadar H, Mariani S (2020) Early damage assessment in large-scale structures by innovative statistical pattern recognition methods based on time series modeling and novelty detection. Adv Eng Softw 150:102923

Entezami A, Shariatmadar H, Sarmadi H (2020) Condition assessment of civil structures for structural health monitoring using supervised learning classification methods. Iran J Sci Technol, Trans Civ Eng 44(1):51–66

Entezami A, Sarmadi H, Salar M, De Michele C, Arslan AN (2020d) A novel data-driven method for structural health monitoring under ambient vibration and high-dimensional features by robust multidimensional scaling. Struct Health Monit

Entezami A, Shariatmadar H, Mariani S (2020e) Structural health monitoring for condition assessment using efficient supervised learning techniques. Proceedings 42(1):17

Farahani RV, Penumadu D (2016) Full-scale bridge damage identification using time series analysis of a dense array of geophones excited by drop weight. Struct Control Health Monit 23(7):982–997

Fassois SD, Sakellariou JS (2009) Statistical time series methods for structural health monitoring. In: Encyclopedia of Structural Health Monitoring. Wiley, pp 443–472

Figueiredo E, Park G, Figueiras J, Farrar C, Worden K (2009) Structural health monitoring algorithm comparisons using standard data sets. Los Alamos National Laboratory: LA-14393

Figueiredo E, Figueiras J, Park G, Farrar CR, Worden K (2011) Influence of the autoregressive model order on damage detection. Comput-Aided Civ Infrastruct Eng 26(3):225–238

Fugate ML, Sohn H, Farrar CR (2001) Vibration-based damage detection using statistical process control. Mech Syst Sig Process 15(4):707–721

Gómez V, Maravall A (2000) Automatic modeling methods for univariate series. In: A course in time series analysis. Wiley, pp 171–201

Gujarati DN, Porter DC (2003) Basic econometrics, 4th edn. McGraw-Hill, New York

Gul M, Necati Catbas F (2009) Statistical pattern recognition for structural health monitoring using time series modeling: theory and experimental verifications. Mech Syst Sig Process 23(7):2192–2204

Gul M, Necati Catbas F (2011) Structural health monitoring and damage assessment using a novel time series analysis methodology with sensor clustering. J Sound Vib 330(6):1196–1210

Hoell S, Omenzetter P (2016) Optimal selection of autoregressive model coefficients for early damage detectability with an application to wind turbine blades. Mech Syst Sig Process 70:557–577

Hyndman RJ, Athanasopoulos G (2014) Forecasting: principles and practice. OTexts

Johnson EA, Lam HF, Katafygiotis LS, Beck JL (2004) Phase I IASC-ASCE structural health monitoring benchmark problem using simulated data. J Eng Mech 130(1):3–15

Kitagawa G (2010) Introduction to time series modeling. CRC Press, Taylor & Francis Group, Boca Raton

Kopsaftopoulos FP, Fassois SD (2015) A vibration model residual-based sequential probability ratio test framework for structural health monitoring. Struct Health Monit 14(4):359–381

Lakshmi K, Rao A, Gopalakrishnan N (2016) Singular spectrum analysis combined with ARMAX model for structural damage detection. Struct Control Health Monit, e1960

Ljung L (1999) System identification: theory for the user, 2nd edn. Prentice-Hall, Upper Saddle River, NJ

Mattson SG, Pandit SM (2006) Statistical moments of autoregressive model residuals for damage localisation. Mech Syst Sig Process 20(3):627–645

Mei L, Mita A, Zhou J (2016) An improved substructural damage detection approach of shear structure based on ARMAX model residual. Struct Control Health Monit 23:218–236

Montgomery DC, Jennings CL, Kulahci M (2015) Introduction to time series analysis and forecasting. Wiley

Mosavi AA, Dickey D, Seracino R, Rizkalla S (2012) Identifying damage locations under ambient vibrations utilizing vector autoregressive models and Mahalanobis distances. Mech Syst Sig Process 26:254–267

Nair KK, Kiremidjian AS, Law KH (2006) Time series-based damage detection and localization algorithm with application to the ASCE benchmark structure. J Sound Vib 291(1):349–368

Rezaiee-Pajand M, Entezami A, Shariatmadar H (2017) An iterative order determination method for time-series modeling in structural health monitoring. Adv Struct Eng 21(2):300–314

Ross SM (2014) Introduction to probability and statistics for engineers and scientists. Academic Press

Roy K, Bhattacharya B, Ray-Chaudhuri S (2015) ARX model-based damage sensitive features for structural damage localization using output-only measurements. J Sound Vib 349:99–122

Saito T, Beck JL (2010) Bayesian model selection for ARX models and its application to structural health monitoring. Earthquake Eng Struct Dynam 39(15):1737–1759

Sarmadi H, Entezami A (2020) Application of supervised learning to validation of damage detection. Arch Appl Mech, in press

Scott DW (2015) Multivariate density estimation: theory, practice, and visualization. Wiley

Sohn H, Farrar CR (2001) Damage diagnosis using time series analysis of vibration signals. Smart Mater Struct 10(3):446–451

Sohn H, Czarnecki JA, Farrar CR (2000) Structural health monitoring using statistical process control. J Struct Eng 126(11):1356–1363

Stoica P, Moses RL (1997) Introduction to spectral analysis, vol 1. Prentice hall Upper Saddle River, NJ

Stull CJ, Hemez FM, Farrar CR (2012) On assessing the robustness of structural health monitoring technologies. Struct Health Monit 11(6):712–723

Yao R, Pakzad SN (2012) Autoregressive statistical pattern recognition algorithms for damage detection in civil structures. Mech Syst Sig Process 31:355–368

Yao R, Pakzad SN (2013) Damage and noise sensitivity evaluation of autoregressive features extracted from structure vibration. Smart Mater Struct 23(2):025007

Young PC (2011) Recursive estimation and time-series analysis: an introduction for the student and practitioner. Springer, Berlin Heidelberg

Zhang QW (2007) Statistical damage identification for bridges using ambient vibration data. Comput Struct 85(7):476–485

Zheng H, Mita A (2008) Damage indicator defined as the distance between ARMA models for structural health monitoring. Struct Control Health Monit 15(7):992–1005

Zheng H, Mita A (2009) Localized damage detection of structures subject to multiple ambient excitations using two distance measures for autoregressive models. Struct Health Monit 8(3):207–222

Chapter 3
Feature Extraction in Time-Frequency Domain for Non-Stationary Data

3.1 Introduction

Although the time-invariant linear models are popular and influential statistical tools for modeling the vibration time-domain measurements, those may not provide reliable consequences of feature extraction when the structural responses are non-stationary. Due to recent advances in sensor technology and dynamic testing, most of the civil engineering structures are normally excited by the ambient vibration sources such as wind, traffic, human activity, etc. Under these kinds of excitations, it is possible to acquire non-stationary vibration responses as slowly trends or seasonality variants from variations in the environment of structures and deliberate operational changes (Worden et al. 2016).

An efficient way of analyzing non-stationary signals is to use time-frequency techniques (Sarmadi et al. 2019; Sarmadi et al. 2020b). Feng et al. (2013) published a review article that comprehensively analyzes the advances in time-frequency methods for mechanical systems. In another review survey, Maheswari and Umamaheswari (2017) evaluated non-stationary signal processing techniques such as linear, non-linear, quadratic time-frequency approaches applied to vibration analysis of a wind turbine. Furthermore, the comprehensive overview of time-frequency and time-scale analysis methods for SHM can be found in the review article of Staszewski and Robertson (2007). One of the most reliable and simple non-stationary signal processing techniques for feature extraction is to exploit adaptive time-frequency data analysis.

For the first time, Huang et al. (1998) proposed the Empirical Mode Decomposition (EMD) technique, which is capable of decomposing a signal into some oscillation modes named as Intrinsic Mode Functions (IMFs). Even though EMD is a beneficial method for analyzing stationary and non-stationary, it suffers from a serious drawback named mode mixing. In order to alleviate this shortcoming, Wu and Huang (2009) developed the Ensemble Empirical Mode Decomposition (EEMD) method by adding white noise to the original signal and calculating using IMFs repeatedly. A comprehensive review of adaptive time-frequency methods on vibration-based applications can be found in the work of Lei et al. (2013). For more details about the applications of EMD and EEMD to SHM, the reader can refer to the research articles of Liu et al. (2006), Pines and Salvino (2006), Chen (2009), Aied et al. (2016), Entezami and Shariatmadar (2019b), Sarmadi et al. (2020a), and (Entezami and Shariatmadar 2019a). Taking the disadvantages of EMD and EEMD into account, Jiang et al. (2013) proposed an improved EEMD with a multi-wavelet packet for the analysis of non-linear and non-stationary vibration signals acquired from mechanical systems for rotating machinery multi-fault diagnosis. To deal with the limitation of EEMD in obtaining redundant IMFs, Colominas et al. (2014) proposed Improved Complete Ensemble Empirical Mode Decomposition with Adaptive Noise (ICEEMDAN).

The ensemble adaptive time-frequency methods depend strongly on two vital parameters including the amplitude of added white noise and ensemble trails or numbers. Zhang et al. (2010) evaluated the performance of EEMD under different amplitude noise amounts and ensemble numbers. They introduced the Signal-to-Noise Ratio (SNR) and correlation measure between the decomposed IMFs and the components in the original signal to provide a systematic approach to choosing an appropriate noise amplitude and a feasible number of ensemble numbers. Guo and Peter (2013) proposed relative Root–Mean–Square–Error (RMSE) between the vibration signal and an IMF, which has the highest correlation with the original signal, to select a proper amplitude white noise. Utilizing the lowest coefficient correlation between two successive IMFs, Kedadouche et al. (2016) presented a new method to select the amplitude of the added white noise for the EEMD technique. The other important issue concerning using the adaptive time-frequency data analysis methods is to select optimal IMFs for damage diagnosis. This issue originates from the fact that the first few IMFs contain the high-frequency components of the original signal, which are more sensitive to the local damage. Hence, it is better to use such IMFs rather than the low-frequency components in damage identification. This is because these components could have not physical meaning and could be caused by the stop criteria set in the sifting process (Ricci and Pennacchi 2011). In this regard, Žvokelj et al. (2011) emphasized on using a few IMFs instead of all IMFs. They applied the kurtosis and crest factor to obtain the highest impulsive nature for choosing the most appropriate IMFs. Ricci and Pennacchi (2011) proposed an index to automatically choose the proper IMFs based on a linear combination of two indices including a measure of the periodicity degree of the IMF and its absolute skewness value. Sarmadi et al. (2020a) proposed a method based on the generalized Pearson correlation function, which measures a complete correlation coefficient based on an accurate selection by considering the internal correlation between data samples.

Sometimes the direct use of adaptive time-frequency data analysis methods may not necessarily be useful for SHM or may not provide sensitive features to damage, particularly under the ambient vibration cases. In such circumstances, hybrid algorithms make reliable and efficient approaches for dealing with the mentioned limitations. The combination of Kernel Principal Component Analysis with EEMD for multivariate vibration signal analysis and bearing fault detection was presented by

Žvokelj et al. (2011). In another study, Guo et al. (2012) developed a hybrid signal processing method that combined Spectral Kurtosis (SK) with EEMD in an effort to detect bearing faults under large noise measurements in non-linear and non-stationary vibration signals. Sharma and Parey (2017) utilized the combination of EMD and Dynamic Time Warping (DTW) to extract efficient features from vibration signals with fluctuating speed conditions. Junsheng et al. (2006) presented a hybrid algorithm for damage detection by combining the EMD with the AR model. With these descriptions, this book aims to proposed new hybrid methods for feature extraction from non-stationary vibration responses. In these methods, all the above-mentioned limitations and challenges are dealt with appropriately.

The proceeding parts of this chapter of the book are as follows: Sect. 3.2 describes some adaptive time-frequency signal decomposition algorithms such as EMD, EEMD, and ICEEMDADN. In Sect. 3.3, a hybrid method as a combination of the ICEEMDAN algorithm and ARMA modeling is proposed to extract features from non-stationary signals. Section 3.4 summarizes the conclusions of this chapter.

3.2 Adaptive Time-Frequency Data Analysis Methods

3.2.1 Empirical Mode Decomposition

The fundamental principle of EMD relies on the decomposition of a signal into IMFs (a set of complete and almost orthogonal components), which are arranged from high to low frequencies based on local characteristic timescales. Each IMF is representative of a natural vibration mode embedded in the signal and serves as a basis function determined by the signal itself. Given the original signal $y(t)$, the EMD algorithm decomposes it into m IMFs as follows (Lei et al. 2013):

$$y(t) = \sum_{i=1}^{m} c_i(t) + r_m(t), \qquad (3.1)$$

where $c_i(t)$ and $r_m(t)$ are the ith IMF and the residual (trend) of the original signal, respectively. Each IMF should satisfy two conditions including (i) the numbers of extreme and zero-crossings should not differ by more than one, and (ii) the mean value of envelope defined by the maxima and minima at any given time should be zero. For the sake of convenience and clarification, the EMD algorithm is summarized as follows:

- **Step 1**—Set $r_0 = y(t)$ as the signal residue at $k = 0$.
- **Step 2**—Find all local minima and maxima of the original signal $y(t)$.
- **Step 3**—Interpolate between minima and maxima by cubic spline lines to obtain the lower and upper envelops (ε_{min} and ε_{max}).
- **Step 4**—Compute the average of envelops $\overline{\varepsilon} = (\varepsilon_{min} + \varepsilon_{max})/2$.

- **Step 5**—Calculate the IMF candidate $c_{k+1} = r_k - \bar{\varepsilon}$.
- **Step 6**—If c_{k+1} is an IMF, save it and compute the residue $r_{k+1} = r_k - c_{k+1}$.
- **Step 7**—Continue until the final residue r_k satisfies some predefined stopping criteria.

3.2.2 Ensemble Empirical Mode Decomposition

The EEMD method adds a uniformly distributed white noise signal with a limited amplitude of the original signal along with an ensemble number to the algorithm of EMD in order to overcome the mode-mixing problem and reconstruct IMFs better than EMD (Wu and Huang 2009; Lei et al. 2013). Since the white noise is added throughout the entire signal decomposition process, no missing scales are present leading to the elimination of mode mixing. By adding the zero mean unit variance white noise $w(t)$ to the original signal, one can write:

$$y_j(t) = y(t) + w_j(t), \quad j = 1, 2, ..., NE, \tag{3.2}$$

where $y_j(t)$ is the noise-added signal and NE denotes the number of ensembles or trails. This scalar amount refers to the number of times that the white noise is added to the signal. On this basis, the noise signal added to the original signal can be decomposed into a series of IMFs in the EEMD method with the aid of the EMD algorithm as follows:

$$y_j(t) = \sum_{i=1}^{m_j} c_{i,j}(t) + r_{m_j}(t), \quad j = 1, 2, ..., NE. \tag{3.3}$$

In Equation (3.3), $c_{i,j}(t)$ represents the ith IMF of the jth ensemble, $r_{mj}(t)$ is the residual of jth ensemble, and m_j denotes the IMFs number of the jth ensemble. By calculating the average of IMFs in all trails, the final IMF $\hat{c}(t)$ obtained by the EEMD algorithm is given by:

$$\hat{c}_i(t) = \frac{\sum_{j=1}^{NE} c_{i,j}(t)}{NE}, \quad i = 1, 2, ..., \hat{m}, \tag{3.4}$$

where $\hat{m} = \min(m_1, m_2, ... m_{NE})$. The algorithm of EEMD method is summarized in the following steps:

- **Step 1**—Initialize the number of ensembles (NE) and the amplitude of white noise (A_n). Please observe Sect. 3.2.4.
- **Step 2**—Generate a zero mean unit variance white noise $w(t)$ by the initialized amplitude and add it to the original signal.

- **Step 3**—Decompose the noise-added signal into m IMFs $c_{i,j}$ using the EMD algorithm as summarized in steps of the previous section.
- **Step 4**—If $j < NE$, then return to step 2 with $j = j + 1$. Once again, repeat steps 2 and 3 using a new white noise.
- **Step 5**—Calculate the ensemble mean $\hat{c}_i(t)$ of the NE trails for each IMF based on Eq. (3.4)
- **Step 6**—Apply the mean $\hat{c}_i(t)$ of each of the $\hat{m}$ IMFs as the final IMF.

3.2.3 Improved Complete Ensemble Empirical Mode Decomposition with Adaptive Noise

In the EEMD algorithm, each noise-added signal is independently decomposed from the other ensembles leading to an individual residue for each of them at each step without any connection between the different ensembles. Under such circumstances, the EEMD algorithm suffers from some disadvantages such as an incomplete decomposition process and different redundant modes (Colominas et al. 2014). Taking into account these drawbacks, the ICEEMDAN method is described here to extract adequate IMFs, which are neither few nor redundant. Before discussing the method, let $\Pi_i(.)$ and $\Omega(.)$ be the operators that produce the ith mode and the local mean of the signal (the average of the upper and lower envelops of the signal interpolated by cubic splines) in the EMD algorithm. Furthermore, let $\langle . \rangle$ be the action of averaging throughout the ensembles. With these definitions, the ICEEMDAN method initially calculates the local mean of NE ensemble, $y_j(t) = y(t) + \Pi_1(w_j(t))$, to obtain the first residue and IMF as follows (Colominas et al. 2014):

$$r_1(t) = \langle \Omega(y_j(t)) \rangle, \tag{3.5}$$

$$c_1(t) = y(t) - r_1(t). \tag{3.6}$$

In the second step, the residue is estimated as the average of the local means of the ensembles $r_1(t) + \Pi_2(w_j(t))$ to define the second mode in the following form (Colominas et al. 2014):

$$c_2(t) = r_1(t) - r_2(t) = r_1(t) - \langle \Omega(r_1(t) + \Pi_2(w_j(t))) \rangle. \tag{3.7}$$

For $i = 3, .., m$, the ith residue is computed as follows (Colominas et al. 2014):

$$r_i(t) = \langle \Omega(r_{i-1}(t) + \Pi_i(w_j(t))) \rangle. \tag{3.8}$$

Eventually, the ith mode is given by:

$$c_i(t) = r_{i-1}(t) - r_i(t). \tag{3.9}$$

The above-mentioned algorithm makes the ICEEMDAN method superior to the EMD and EEMD techniques due to dealing with the mode mixing, avoiding redundant or spurious modes, and decreasing the amount of noise available in the IMFs.

3.2.4 Selection of Noise Amplitude and Ensemble Number

The performance of each noised-assisted signal decomposition methods (e.g. EEMD, ICEEMDAN, etc.) depends strongly on choosing an accurate noise amplitude (A_n) and a sufficient ensemble number (Guo and Peter 2013). In most cases, these values are obtained from trial and error or empirical equations. For instance, Wu and Huang (2009) provided a relationship among the ensemble number, the noise amplitude, and the standard deviation of error as follows:

$$\ln \sigma_e + \frac{A_n}{2} \cdot \ln NE = 0, \tag{3.10}$$

where σ_e represents the standard deviation of error between the original signal and the corresponding IMF. They recommended that the amplitude of the added white noise is approximately 0.2 or 20% of the standard deviation of the original signal and the value of ensemble number is a few hundred. However, this approach is not always useful for signal processing in various applications, because the noise amplitude is of paramount importance to the performance of the noised-assisted signal decomposition methods. A very low value of A_n will not introduce adequate changes in the extremes of the decomposed signal. In contrast, the selection of a very high noise amplitude will result in redundant IMFs. In order to have an efficient choice of A_n, one can introduce relative root-mean-square of error (R_{RMSE}) method proposed by Guo and Peter (2013) as follows:

$$R_{RMSE} = \sqrt{\frac{\sum_{t=1}^{n} (y(t) - c_{\max}(t))^2}{\sum_{t=1}^{n} (y(t) - \bar{y})^2}}, \tag{3.11}$$

where $c_{\max}$ is the main IMF that has the highest correlation with the original signal. Furthermore, $\bar{y}$ and n represent the mean and the number of data samples of the original signal, respectively. If the value of R_{RMSE} becomes small, the selected IMF is close to the original signal with the availability of white noise. This means that $c_{\max}$ not only contains the main component of the original signal but also includes the noise and/or the other irrelevant signal components. On this basis, the difference between the original signal and the selected IMF becomes small implying an inappropriate decomposition process. On the contrary, a high value of R_{RMSE} indicates that the signal is separated from the noise and the other irrelevant signal components. To put it another way, this R_{RMSE} value confirms a proper decomposition result so that the selected IMF consists of the main signal component. It is worth remarking that the amplitude of the added white noise is related to the original signal and expressed as (Guo and Peter 2013):

$$A_n = L_n \sigma_y, \tag{3.12}$$

where σ_y is the standard deviation of the original signal and L_n denotes the noise level of the added white noise. Since σ_y is usually constant, the selection of amplitude noise is equivalent to choosing the noise level L_n.

Once the noise amplitude has been obtained, it is necessary to determine an ensemble number. Note that this process may contain two major limitations. First, the selection of a large ensemble amount will lead to a higher computational cost. Second, a small number of NE will not enable the noise-assisted signal decomposition methods to cancel out the noise remaining in each IMF (Guo and Peter 2013). Regarding the optimal selection of A_n, the signal-to-noise ratio (SNR) can be applied to determine the appropriate ensemble number. This process is based on fixing the optimal noise amplitude and increasing the ensemble number until the change in the value of SNR relatively becomes small. With these descriptions, in this work, one attempts to choose the noise amplitude and the ensemble number via the above-mentioned equations and approaches.

3.3 A New Hybrid Feature Extraction Method by ICEEMDAN and ARMA

3.3.1 A Novel Automatic IMF Selection Approach

Even though the ICEEMDAN method reduces the number of redundant IMFs, it is important to choose relevant IMFs to structural changes caused by damage. In general, the first few IMFs contain the high-frequency components of the original signal that are more sensitive to the local damage. Hence, it is better to use such IMFs rather than the low-frequency components in damage identification. This is because these components, which are extracted from at the end of the sifting process, may not have physical meaning (Ricci and Pennacchi 2011). On the other hand, the choice of sensitive IMFs is usually carried out by the user experience and visual inspection of the IMF plots. A novel automatic IMF selection approach is proposed to select the most relevant IMFs based on their energy levels and a mathematical function named as mode participation factor (MPF). The total energy of all IMFs (E_T) is given by:

$$E_T = \sum_{i=1}^{m} E_{\hat{c}_i(t)},$$

(3.13)

where $E_{\hat{c}_i(t)}$ denotes the amount of energy contained in the ith IMF, which can be expressed as:

$$E_{\hat{c}_i(t)} = \sum_{t=1}^{n} |\hat{c}_i(t)|^2.$$

(3.14)

Based on the Eqs. (3.13) and (3.14), the MPF of each IMF is presented as follows:

$$MPF_i(\%) = \frac{E_{\hat{c}_i(t)}}{E_T} \times 100, \qquad i = 1, 2, \ldots, m.$$

(3.15)

The central idea behind the proposed approach is to select IMFs that the sum of their MPFs is more than a predefined value (β). To achieve this goal, all obtained MPFs should be arranged in descending order. As a result, the optimal number of IMFs is automatically obtained in the following form:

$$\sum_{i=1}^{m} MPF_i \geq \beta \quad \rightarrow \quad m = m_{opt}.$$

(3.16)

It is significant to mention that the main idea of the proposed IMF selection approach originates from the theory of the modal participation factor regarding the structural dynamics and modal analysis (Paultre 2011). Inspired by this theory, one can define $\beta = 90\%$ to choose sufficient and optimal IMFs automatically.

3.3.2   Proposed Hybrid Algorithm

The proposed hybrid approach is a two-stage algorithm that it is established by combining the ICEEMDAN method and ARMA model. The main objective of this algorithm is to extract the residuals of ARMA models fitted to the selected IMFs as the main DSFs of the undamaged and damaged states. In the first stage, a non-stationary and/or stationary vibration signal is initially decomposed into several IMFs using the ICEEMDAN method.

The second stage is concerned with fitting an individual ARMA model to each selected IMF. In this stage, time series modeling by ARMA includes the determination of the model orders (p and q) and estimation of its coefficients ($\Theta_u = [\theta_{u_1} \ldots \theta_{u_p}]$ and $\Psi_u = [\psi_{u_1} \ldots \psi_{u_q}]$) from the normal condition of the structure and then the extraction of the ARMA residuals of both the undamaged and damaged states. Assume that $c_{mu}(t)$ and $c_{md}(t)$ are one of the selected IMFs extracted from the proposed automatic IMF selection approach in the normal and damaged structures, respectively. The model residuals as feature vectors of these conditions are extracted in the following forms:

$$e_u(t) = c_{mu}(t) - \left(\sum_{i=1}^{p} \theta_{u_i} c_{mu}(t-i) + \sum_{j=1}^{q} \psi_{u_j} e_u(t-j) \right), \qquad (3.17)$$

$$e_d(t) = c_{md}(t) - \left(\sum_{i=1}^{p} \theta_{u_i} c_{md}(t-i) + \sum_{j=1}^{q} \psi_{u_j} e_d(t-j) \right). \qquad (3.18)$$

Finally, the model residuals from each IMF, and each sensor are extracted and used as the main DSFs. These features are then applied to distance measures that are proposed in the next section in order to detect and locate damage.

3.4 Conclusions

This chapter of the book has concentrated on the process of feature extraction for the non-stationary vibration signals. It has been mentioned that the conventional signal processing techniques in time and frequency domains are normally proper for stationary data. Hence, those may not provide reliable features from the non-stationary signals. To deal with this problem, some adaptive time-frequency data analysis methods such as EMD, EEMD, and ICEEMDAN have been presented. It has been pointed out that the direct use of IMFs obtained from such techniques may fail in extracting sensitive features to damage when the measured vibration signals are caused by the unmeasurable ambient vibration. On this basis, ICEEMDAN-ARMA approach has been proposed in this chapter of this book to overcome the mentioned limitation.

References

Aied H, González A, Cantero D (2016) Identification of sudden stiffness changes in the acceleration response of a bridge to moving loads using ensemble empirical mode decomposition. Mech Syst Sig Process 66:314–338

Chen J (2009) Application of empirical mode decomposition in structural health monitoring: Some experience. Adv Adapt Data Anal 1(04):601–621

Colominas MA, Schlotthauer G, Torres ME (2014) Improved complete ensemble EMD: a suitable tool for biomedical signal processing. Biomed Sig Process Control 14:19–29

Entezami A, Shariatmadar H (2019) Damage localization under ambient excitations and non-stationary vibration signals by a new hybrid algorithm for feature extraction and multivariate distance correlation methods. Struct Health Monit 18(2):347–375

Entezami A, Shariatmadar H (2019) Structural health monitoring by a new hybrid feature extraction and dynamic time warping methods under ambient vibration and non-stationary signals. Measurement 134:548–568

Feng Z, Liang M, Chu F (2013) Recent advances in time–frequency analysis methods for machinery fault diagnosis: a review with application examples. Mech Syst Sig Process 38(1):165–205

Guo W, Peter WT (2013) A novel signal compression method based on optimal ensemble empirical mode decomposition for bearing vibration signals. J Sound Vib 332(2):423–441

Guo W, Peter WT, Djordjevich A (2012) Faulty bearing signal recovery from large noise using a hybrid method based on spectral kurtosis and ensemble empirical mode decomposition. Measurement 45(5):1308–1322

Huang NE, Shen Z, Long SR, Wu MC, Shih HH, Zheng Q, Yen N-C, Tung CC, Liu HH (1998) The empirical mode decomposition and the Hilbert spectrum for nonlinear and non-stationary time series analysis. Proc R Soc Lond A 454:903–995, The Royal Society, vol 1971. The Royal Society, pp 903–995

Jiang H, Li C, Li H (2013) An improved EEMD with multiwavelet packet for rotating machinery multi-fault diagnosis. Mech Syst Sig Process 36(2):225–239

Junsheng C, Dejie Y, Yu Y (2006) A fault diagnosis approach for roller bearings based on EMD method and AR model. Mech Syst Sig Process 20(2):350–362

Kedadouche M, Thomas M, Tahan A (2016) A comparative study between empirical wavelet transforms and empirical mode decomposition methods: application to bearing defect diagnosis. Mech Syst Sig Process 81:88–107

Lei Y, Lin J, He Z, Zuo MJ (2013) A review on empirical mode decomposition in fault diagnosis of rotating machinery. Mech Syst Sig Process 35(1):108–126

Liu J, Wang X, Yuan S, Li G (2006) On Hilbert-Huang transform approach for structural health monitoring. J Intell Mater Syst Struct 17(8):721–728

Maheswari RU, Umamaheswari R (2017) Trends in non-stationary signal processing techniques applied to vibration analysis of wind turbine drive train–a contemporary survey. Mech Syst Sig Process 85:296–311

Paultre P (2011) Dynamics of structures. Wiley

Pines D, Salvino L (2006) Structural health monitoring using empirical mode decomposition and the Hilbert phase. J Sound Vib 294(1):97–124

Ricci R, Pennacchi P (2011) Diagnostics of gear faults based on EMD and automatic selection of intrinsic mode functions. Mech Syst Sig Process 25(3):821–838

Sarmadi H, Entezami A, Daneshvar Khorram M (2019) Energy-based damage localization under ambient vibration and non-stationary signals by ensemble empirical mode decomposition and Mahalanobis-squared distance. J Vib Control 26(11–12):1012–1027

Sarmadi H, Entezami A, Daneshvar Khorram M (2020) Energy-based damage localization under ambient vibration and non-stationary signals by ensemble empirical mode decomposition and Mahalanobis-squared distance. J Vib Control 26(11–12):1012–1027

Sarmadi H, Entezami A, Saeedi Razavi B, Yuen K-V (2020c) Ensemble learning-based structural health monitoring by Mahalanobis distance metrics. Struct Control Health Monit:e2663

Sharma V, Parey A (2017) Frequency domain averaging based experimental evaluation of gear fault without tachometer for fluctuating speed conditions. Mech Syst Sig Process 85:278–295

Staszewski WJ, Robertson AN (2007) Time–frequency and time–scale analyses for structural health monitoring. Philos Trans Royal Soc London A: Math, Phys Eng Sci 365(1851):449–477

Worden K, Baldacchino T, Rowson J, Cross EJ (2016) Some Recent Developments in SHM Based on Nonstationary Time Series Analysis. Proc IEEE 104(8):1589–1603

Wu Z, Huang NE (2009) Ensemble empirical mode decomposition: a noise-assisted data analysis method. Adv Adapt Data Anal 1(1):1–41

Zhang J, Yan R, Gao RX, Feng Z (2010) Performance enhancement of ensemble empirical mode decomposition. Mech Syst Sig Process 24(7):2104–2123

Žvokelj M, Zupan S, Prebil I (2011) Non-linear multivariate and multiscale monitoring and signal denoising strategy using kernel principal component analysis combined with ensemble empirical mode decomposition method. Mech Syst Sig Process 25(7):2631–2653

Chapter 4
Statistical Decision-Making by Distance Measures

4.1 Introduction

Novelty detection is an unsupervised learning or one-class classification technique that mainly aims at learning a statistical model or detector by training data and detecting any abnormality, anomaly, outlier, and damage by using testing data (Pimentel et al. 2014). In the context of SHM, the training data is obtained from the DSFs extracted from the vibration responses of the normal conditions, which often include different environmental and/or operational variations. Furthermore, the testing data is usually constructed from a small part of DSFs regarding the normal conditions and all DSFs of the current state of the structure. It is important to note that the procedures of constructing the training data and learning the statistical model are carried out in the training or baseline phase. Additionally, the steps of constructing the testing data and decision-making for SHM is performed in the monitoring or inspection stage. The novelty detection can be defined as the method of recognizing that a sample in the testing data differs in some respect from the training data. To put it another way, the process of novelty detection refers to the problem of finding anomalies or outliers in the testing data that do not conform to the statistical model or classifier learned by the training data (Markou and Singh 2003).

In some literature, novelty detection is known as anomaly detection or outlier detection. The different terms originate from diverse domains of applications. Nevertheless, the term "*anomaly detection*" is typically applied synonymously with "*novelty detection*". Based on the definition of Barnett and Lewis (1994), an outlier is a data point that appears to be inconsistent with the remainder of data. Therefore, the outlier detection aims at finding this point in a set of data, which can have a large effect on the analysis of the data. In other words, outliers are assumed to contaminate the dataset under consideration. Hence, the objective of outlier detection is to filter out their presence or effects during the training period. On the other hand, the anomalies are patterns in data that are not compatible with the remaining parts of data. These patterns are sufficiently far away from the regions of normality due to irregularities and/or transient events. From this definition, the main goal of anomaly detection is to learn a model of normality from a set of data that is considered normal, and a decision process on the testing data is based upon the use of this classifier (Pimentel et al. 2014). This means that novelty detection slightly differs from the anomaly detection and outlier detection. As Markou and Singh (2003) stated, a novel data point is a new or unknown pattern in the testing data that one cannot find any information about it during the training phase. Based on this definition, one can express that the novelty detection is a process of identification (detection) of an unknown pattern in the testing dataset with the aid of the classifier learned by the training data. This definition of novelty detection is highly similar to the levels of damage diagnosis in SHM because there is no information about damage in the current state of the structure. Therefore, one can realize that the damage diagnosis under the unsupervised learning manner falls into the definition of novelty detection.

For the implementation of novelty detection in SHM, the DSFs extracted from the measured vibration responses of the normal conditions of the structure are collected as the training data. Using them, a statistical model or classifier is learned in an unsupervised learning manner, which is more practical than unsupervised learning strategies (Entezami et al. 2020d; Sarmadi and Entezami 2020) since no information of the damaged state is available. This process is carried out in the training (baseline) phase, in which all of the structural conditions are supposed to be normal along with different EOV. By extracting the same DSFs from the vibration measurements of the current or damaged condition, these are collected as the testing datasets. In some cases, a few or all samples of training data are added to the testing set in order to increase the reliability of decision-making. This procedure is implemented in the monitoring (inspection) phase, in which it is evaluated whether the extracted features of the current state are far away from the learned classifier or not. Without consideration of the EOV or removal of their effects, any deviation of the DSFs of this state from the classifier is indicative of damage occurrence (Farrar and Worden 2013).

The proceeding parts of this chapter of the book are as follows: Sect. 4.2 briefly describes the concepts of statistical distance measures. In Sect. 4.3, some conventional distance measures are explained and formulated. Section 4.4 proposes univariate distance methods for damage localization. In Sect. 4.5, two multivariate distances are proposed to use in the first level of SHM and damage diagnosis; that is, damage detection. Section 4.6 presents a hybrid distance-based novelty detection method for damage detection. In Sect. 4.7, the proposed multi-level distance method is presented and explained. The concepts of two well-known threshold estimation methods are described in Sect. 4.8. Finally, Sect. 4.9 summarizes the conclusions of this chapter.

4.2 Statistical Distance Measure

In statistics, a distance measures the similarity or dissimilarity between two sets of objects, which can be two random variables or two probability distributions. A statistical distance between the probability distributions can be interpreted as measuring the similarity between probability measures. The distance measures associated with the differences between random variables often incorporate correlation in data rather than using their random amounts. This means that those pertain to finding the extent of dependence between random variables and differ from distance measures of probability distributions (Deza and Deza 2014; Aggarwal 2016).

On the other hand, the statistical distance methods are generally either univariate or multivariate. A univariate distance measures the similarity between two sets of vectors, whereas a multivariate distance refers to measuring the distance between two sets of matrices. Some statistical distance measures are not mainly metrics; that is, those do not need to be symmetric. These types of distance measures are referred to as statistical divergences (Deza and Deza 2014).

4.3 Conventional Distance Approaches

4.3.1 General Problem

Assume that $\mathbf{x}, \mathbf{z} \in \Re^{nr}$ are the vectors and $\mathbf{X}, \mathbf{Z} \in \Re^{nr \times nc}$ are the matrices of features. According to the principle of machine learning, $\mathbf{x}$, $\mathbf{X}$, and $\mathbf{z}$, $\mathbf{Z}$ refer to the training and testing datasets. Furthermore, one supposes that $d(\mathbf{z}\|\mathbf{x})$ and $D(\mathbf{Z}\|\mathbf{X})$ refer to the symbols of the univariate and multivariate distances, which are scalar values. Each of them may have the following properties (Deza and Deza 2014):

$$d(\mathbf{z} \,\|\, \mathbf{x}) = 0 \quad \text{or} \quad D(\mathbf{Z} \,\|\, \mathbf{X}) = 0, \tag{4.1}$$

$$d(\mathbf{x} \,\|\, \mathbf{x}) = d(\mathbf{z} \,\|\, \mathbf{z}) = 0 \quad \text{or} \quad D(\mathbf{X} \,\|\, \mathbf{X}) = D(\mathbf{Z} \,\|\, \mathbf{Z}) = 0, \tag{4.2}$$

$$d(\mathbf{z} \,\|\, \mathbf{x}) > 0 \quad \text{or} \quad D(\mathbf{Z} \,\|\, \mathbf{X}) > 0, \tag{4.3}$$

$$d(\mathbf{x} \,\|\, \mathbf{z}) = d(\mathbf{z} \,\|\, \mathbf{x}) \quad \text{or} \quad D(\mathbf{X} \,\|\, \mathbf{Z}) = D(\mathbf{Z} \,\|\, \mathbf{X}). \tag{4.4}$$

Based on Eq. (4.1), one can argue that the vectors or matrices of features are similar; that is, $\mathbf{x} = \mathbf{z}$ and $\mathbf{X} = \mathbf{Z}$. This is a simple interpretation of the similarity between two sets of objects. The property in Eq. (4.2) is always valid for all statistical distances. Moreover, Eqs. (4.3) and (4.4) represent the non-negativity and symmetry of the univariate and multivariate distances. Note that the property in Eq. (4.4) refers to the fact that $d(\mathbf{z}\|\mathbf{x})$ and $D(\mathbf{Z}\|\mathbf{X})$ are metrics. In other words, statistical divergences do not satisfy this property, in which case $d(\mathbf{z}\|\mathbf{x})$ and $D(\mathbf{Z}\|\mathbf{X})$ are not equal to $d(\mathbf{x}\|\mathbf{z})$ and $D(\mathbf{X}\|\mathbf{Z})$.

4.3.2 Kullback–Leibler Divergence

The KLD technique is one of the univariate distance approaches proposed by Kullback and Leibler (1951). It is a non-symmetric measure of the difference between two probability distributions or probability density functions (PDF). KLD attempts to measure how a probability distribution or PDF diverges from the other one. Assuming that $\mathbf{P_x}$ and $\mathbf{P_z}$ are the PDFs of the random feature vectors $\mathbf{x}$ and $\mathbf{z}$, the general formulation of KLD is expressed as:

$$d_{KLD}(\mathbf{P_x}\|\mathbf{P_z}) = \sum_l \mathbf{P_x}(l) \log \frac{\mathbf{P_x}(l)}{\mathbf{P_z}(l)}, \qquad l = 1, 2, ..., nr. \tag{4.5}$$

Since $\mathbf{P_x}$ and $\mathbf{P_z}$ are positive, d_{KLD} always gives a positive distance or divergence amount. In a simple case, $d_{KLD} = 0$ means that two probability distributions have the same behavior and any deviation from zero is representative of dissimilarity between them.

4.3.3 Mahalanobis-Squared Distance

The MSD is a powerful and well-known statistical distance metric that aims to measure the similarity between two multivariate datasets. This method does not depend on the scale or amount of the variables and performs the similarity computation based on the correlation between variables (Sarmadi and Karamodin 2020; Sarmadi et al. 2020a; Deraemaeker and Worden 2018; Nguyen et al. 2014b; Yeager et al. 2019; Entezami et al. 2019b; Entezami et al. 2018; Sarmadi et al. 2020b; Entezami et al. 2020b). Suppose that $\mathbf{N} \in \Re^{nc}$ and $\mathbf{\Sigma} \in \Re^{nc \times nc}$ are the mean vector and covariance matrix of the multivariate feature $\mathbf{X} \in \Re^{nr \times nc}$, where $nr > nc$. The MSD calculates the distance of each row vector of $\mathbf{Z}$ as follows:

$$d_{MSD}(l) = (\mathbf{z}_l - \bar{\mathbf{x}}) \mathbf{\Sigma}^{-1} (\mathbf{z}_l - \bar{\mathbf{x}})^T, \qquad l = 1, 2, ..., nr^*, \tag{4.6}$$

where nr^* is the rows of the matrix $\mathbf{Z}$, which can be $nr^* = nr$, $nr^* > nr$, and $nr^* < nr$. By measuring all nr^* distance values, the similarity or dissimilarity between the $\mathbf{X}$ and $\mathbf{Z}$ is obtained by:

$$\mathbf{D}_{MSD}(\mathbf{Z}\|\mathbf{X}) = \left[d_{MSD}(1) \ d_{MSD}(2) \ \cdots \ d_{MSD}(nr^*) \right]^T, \tag{4.7}$$

in which, $\mathbf{D}_{MSD}$ is a vector. Additionally, both $\mathbf{Z}$ and $\mathbf{X}$ should have the same number of variables (nc). The fundamental principles of MSD lie in some important issues. First, the MSD takes into account the correlation in the data, since it is calculated using the inverse of the variance–covariance matrix of the dataset. In this regard, if $nr = nc$ or $nr < nc$, the MSD fails in measuring accurate distance values between the multivariate features. When $nr < nc$, this implies that $\mathbf{X}$ contains much redundant or correlated information. This is called multicollinearity in the data that leads to a singular or nearly singular variance–covariance matrix in the sense that $\mathbf{\Sigma}$ is invertible (De Maesschalck et al. 2000). Second, the MSD technique is based on the assumption that the distributions of the multivariate features should be normal or Gaussian. This is due to the use of sample mean and covariance for estimating $\bar{x}$ and $\mathbf{\Sigma}$, which are typically useful for the normal distributions (Nguyen et al. 2014a).

The central idea behind the use of MSD for damage diagnosis lies in the fact that if the training dataset ($\mathbf{X}$) satisfies all the above-mentioned requirements and it covers a broad range of EOV conditions in the training phase, the feature vector of the damaged structure with the sources of EOV will deviate from the mean of the normal condition. On the contrary, if the feature vector comes from the undamaged state, even under varying operational and environmental conditions, it will be close to the mean of the normal condition indicating that the structure is undamaged (Figueiredo and Cross 2013).

4.3.4 Kolmogorov–Smirnov Test Statistic

The KSTS is a non-parametric statistical hypothesis test that evaluates the cumulative distribution functions (CDFs) of two sample data vectors in an effort to find whether their CDFs come from the same distributions (Wang et al. 2003). In addition, this hypothesis test presents an opportunity to measure the distance between two CDFs. The main benefit of KSTS against some hypothesis tests is that no prior assumption towards the distribution of two sample data vectors ($\mathbf{x}$ and $\mathbf{z}$) is needed. The Kolmogorov–Smirnov test statistic is given by:

$$d_{KSTS}(\mathbf{z}||\mathbf{x}) = \max(|\mathbf{F_x} - \mathbf{F_z}|), \tag{4.8}$$

where $\mathbf{F_x}$ and $\mathbf{F_z}$ are the empirical CDFs of the feature vectors $\mathbf{x}$ and $\mathbf{z}$, respectively. Because the empirical CDF approximates the distribution of data, one can utilize it instead of the CDF into Eq. (4.8). For each of the feature vectors $\mathbf{x}$ and $\mathbf{z}$, the empirical CDF is expressed as (van der Vaart 2000):

$$\mathbf{F} = \frac{1}{nr} \sum_{k=1}^{nr} \Gamma(k \leq nr), \tag{4.9}$$

in which Γ is an indicator of the probability event corresponding to one for k < *nr*; otherwise, it is identical to zero. In this research work, the conventional distance approaches are applied to compare with the proposed distance measures that are described in the next section.

4.4 Proposed Univariate Distance Methods

4.4.1 *KLD with Empirical Probability Measure*

Although the distance methods are effective and efficient statistical tools for measuring the similarity or discrepancy between two sets of samples, it is often difficult to introduce a robust statistical distance method for measuring the similarity of the time series samples with high dissimilarity (damage) detectability. This is because time series are essentially high-dimensional data and their direct use may lead to a time-consuming and complicated decision-making process (Wang et al. 2013b). To overcome this limitation, Kullback–Leibler Divergence with Empirical Probability Measure (KLDEPM) has been proposed to develop a univariate similarity measure that does not need to compute the probability distribution of data. The key components of this method are the segmentation of random samples and the application of empirical probability measure (Wang et al. 2005). Moreover, the great benefit of this method is to measure the similarity of high-dimensional time series data sets. For this purpose, it exploits a data-partitioning algorithm for dividing random sequences into independent partitions. On this basis, it only utilizes the number of samples within each partition in the distance computation rather than applying the original random values. Furthermore, this algorithm not only enables the KLDEPM method to address the major limitation of using the high-dimensional data but also allows it to treat as a dimensionality reduction approach. As the other advantage, this method is capable of computing the distance of both correlated and uncorrelated random sets. In this research work, the KLDEPM method is employed to locate structural damage by using the residual vectors of time series models for the undamaged and damaged conditions ($\mathbf{e_u}$ and $\mathbf{e_d}$).

Assume that both the residual vectors consist of n samples. Partitioning of the random variables in the KLDEPM method is based upon the maximum entropy approach. To implement this process, the residual vector of the damaged or current state is arranged in ascending order in such a way that it begins with the minimum residual sequence and ends with the maximum one. Accordingly, the arranged vector $\mathbf{e_d}$ is divided into c partitions $\mathbf{H1,H2},\ldots,\mathbf{H}_c$ through the maximum entropy approach in the following form:

$$e_d|_{\min} \leq \mathbf{H}_1 \leq e_d|_{N_d}$$
$$e_d|_{(r-1)N_d} < \mathbf{H}_r \leq e_d|_{(rN_d)}, \qquad (4.10)$$
$$e_d|_{N_d(c-1)} < \mathbf{H}_c \leq e_d|_{\max}$$

where r = 2,3,...,c-1. In this equation, N_d denotes the number of samples in each partition of $\mathbf{e_d}$, which is empirically expressed as $N_d = \sqrt{n}$ (Wang et al. 2005). In such a case, the number of partitions is given by $c = n/N_d$. It is important to mention that N_d and c should be positive integers. For the positive non-integers (decimal), one needs to round them off to the nearest integer less than or equal to themselves. Given a random dataset $\mathbf{A} = [a_1, a_2, ..., a_n]$ partitioned into c segments $\mathbf{H1, H2, ..., H_c}$, the empirical probability measure P_a for the segment $\mathbf{H_r}$ is given by (Dudley 1978):

$$P_a(\mathbf{H}_r) = \frac{1}{n} \sum_{j=1}^{n} \delta_a, \qquad (4.11)$$

where r = 1,2,...,c and $\delta_a = 1$ if and only if $a_j \in \mathbf{H}_r$; otherwise, $\delta_a = 0$. The output of Eq. (4.11) is always a positive real number (either integer or non-integer), which indicates how many samples of $\mathbf{A}$ fall into the domain of $\mathbf{H}_r$. Accordingly, the sum of all values of δ_a equal to 1 concerning the total number of samples in $\mathbf{A}$ represents the empirical probability measure. For example, suppose that $\mathbf{A} = [1,2,...,30]$, which is partitioned into the three partitions $\mathbf{H}_1 = [1,2,...10]$, $\mathbf{H}_2 = [11,12,...,20]$, and $\mathbf{H3} = [21,22,...,30]$. Figure 4.1 shows the empirical probability measures of these partitions.

As can be seen in Fig. 4.1, it is clear that the empirical probability measure for each segment is identical to 10/30 or 1/3. Although this process is straightforward, the calculation of empirical probability measure depends directly on the number of segments and data samples within each segment. As mentioned above, the strategy of segmentation used in the proposed KLDEPM method is based on the maximum entropy approach. Let P_u and P_d be the empirical probability measures of the residual

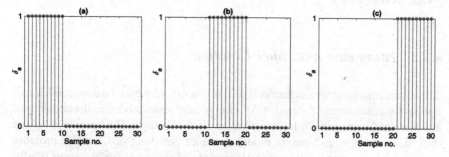

Fig. 4.1 Empirical probability measures: **a** H_1, **b** H_2, **c** H3

vectors $\mathbf{e_u}$ and $\mathbf{e_d}$. By utilizing the numerical information of the partitioning process (i.e. N_d, N_u, and c), these measures for the rth partition are given by:

$$P_u(\mathbf{H}_r) = \frac{N_{u_r}}{n},$$ (4.12)

$$P_d\left(\mathbf{H}_{r,r\neq c}\right) = \frac{N_d}{n},$$ (4.13)

where N_{u_r} denotes the number of samples of $\mathbf{e_u}$ within the segment $\mathbf{H}_r$. It is worth remarking that the first $c - 1$ partitions of the residual vector $\mathbf{e_d}$ have the same number of samples except for the cth partition. For this partition, the empirical probability measure is written as follows:

$$P_d(\mathbf{H}_c) = \frac{n - N_d c}{n}.$$ (4.14)

Once the empirical probability measures have been determined, the KLDEPM method is formulated as:

$$d_{KLDEPM}(\mathbf{e_u}\|\mathbf{e_d}) = \sum_{r=1}^{c} P_u(\mathbf{H}_r) \log \frac{P_u(\mathbf{H}_r)}{P_d(\mathbf{H}_r)}.$$ (4.15)

It is clear from Eqs. (4.12)–(4.15) that the proposed KLDEPM formulated in Eq. (4.15) calculates the distance between two randomly datasets by the numerical information of partitioning and empirical probability measures without directly utilizing the random samples in the distance calculation and obtaining their probability distributions similar to the classical KLD technique. Since all numerical amounts gained by the process of segmentation are positive, $d_{KLEPM} \geq 0$. Ideally, a zero-distance quantity denotes the similarity between the residual vectors of the normal and current structural states, which means that no difference (damage) is present between them.

4.4.2 Parametric Assurance Criterion

The Parametric assurance criterion (PAC) is a coefficient-based damage index that utilizes the coefficients of output or AR term of time series models in the undamaged and damaged conditions to locate and quantify the damage. The basic idea behind the PAC method comes from the modal assurance criterion (MAC), which provides a measure of consistency between experimental and analytical modal vectors (Pastor et al. 2012). The MAC takes on values from zero to one, in which the zero is indicative of no consistency and one represents an entire consistency between the modal vectors.

For the problem of damage diagnosis, if the MAC value corresponds to 1, this means that there is no damage to the structure, and zero otherwise (Baghiee et al. 2009). In this research, the basic concept of MAC is used to establish a new damage index for locating damage and estimating its quantity. Suppose that Θ_u and Θ_d are the vectors of the AR coefficients in the undamaged and damaged conditions extracted from the CBFE approach. Accordingly, the proposed PAC method is formulated as follows:

$$PAC = \frac{\left(\Theta_u^T \cdot \Theta_d\right)^2}{\left(\Theta_u^T \cdot \Theta_u\right)\left(\Theta_d^T \cdot \Theta_d\right)}. \tag{4.16}$$

By computing the PAC values at all sensors, a vector with ns elements $\mathbf{d_P} = [PAC_1 \ldots PAC_{ns}]$ can be achieved, where ns denotes the number of sensors mounted on the structure. In a similar way to the MAC, the values of PAC vary from zero to one ($0 \leq PAC \leq 1$), in which the sensor locations with the PAC quantities close to one and zero are representative of the undamaged and damaged areas of the structure.

4.4.3 Residual Reliability Criterion

The Residual reliability criterion (RRC) is a residual-based damage index that is established by the theory of reliability index. In structural reliability analysis, this index is a useful indicator for computing the failure probability (Nowak and Collins 2012). For the normal distributed random variables, the reliability index is simply determined as the ratio of the mean to standard deviation; that is, $\beta = \mu/\sigma$. The key concept of RRC is to calculate the relative error between the reliability indices of the undamaged and damaged conditions by using the mean and standard deviation of the residuals of time series models extracted from the RBFE approach ($\mathbf{e_u}$ and $\mathbf{e_d}$). Suppose that μ_u and μ_d refer to the mean of $\mathbf{e_u}$ and $\mathbf{e_d}$. In addition, σ_u and σ_d denote the standard deviations of the model residuals for the undamaged and damaged states. As a simple form, the relative reliability index (β_r) between the reliability indices of the undamaged and damaged states is written as follows:

$$\beta_r = \frac{\beta_d - \beta_u}{\beta_u}. \tag{4.17}$$

Based on Eq. (4.17), it is possible to encounter negative values of the relative reliability index because the mean value of a random variable may be a negative amount. To deal with this issue, both numerator and denominator of Eq. (4.17) are squared to develop β_r in the following form:

$$\beta_r = \frac{(\beta_d - \beta_u)^2}{\beta_u^2}. \tag{4.18}$$

By inserting the mean and standard deviation of the residuals in the undamaged and damaged conditions into Eq. (4.18), the main formulation of RRC is proposed as:

$$RRC = \frac{\left(\frac{\mu_d}{\sigma_d} - \frac{\mu_u}{\sigma_u}\right)^2}{\left(\frac{\mu_u}{\sigma_u}\right)^2} = \frac{\mu_d^2 \sigma_u^2 - 2\mu_u \mu_d \sigma_d \sigma_u + \mu_u^2 \sigma_d^2}{\mu_u^2 \sigma_d^2}. \qquad (4.19)$$

From this equation, one observes that the RRC method is only formulated by the statistical moments of the residual vectors e_u and e_d. By computing the RRC values at all sensors, a vector with ns elements $d_R = [RRC_1 \ldots RRC_{ns}]$ is obtained, where each element of this vector denotes the RRC quantity at each sensor location.

Similarly, with the PAC method, the values of RRC are in the range of zero to one ($0 \leq RRC \leq 1$), with the difference that the sensor location with the RRC value close to zero is representative of the undamaged area of the structure, whereas the sensor location with the RRC near to one implies the damaged area or the location of the damage.

4.5   Proposed Multivariate Distance Methods

The correlation coefficient is an efficient statistical tool for measuring correlation as a linear relationship between two random variables. In signal processing, the correlation coefficient between two signals in time or frequency domains measures their similarity or discrepancy. Although there are many applications of the correlation coefficient to SHM (Catbas et al. 2012; Mustapha et al. 2014; Wang et al. 2013a; Wang et al. 2009), the great limitation of using this statistical tool is that the random variables (either vectors or matrices) should have the same dimension. On the other hand, it is essential to use methods that are efficiently and reliably capable of measuring the similarity between two random high-dimensional multivariate data sets.

Having considered these issues, sample distance correlation (SDC) and modified distance correlation (MDC) methods are presented to calculate the similarity of multivariate high-dimensional datasets as the residual matrices of time series models from the undamaged and damaged conditions (E_u and E_d). These methods characterize the correlation between multivariate random variables in an arbitrary dimension. The distance correlation approaches are analogous to product-moment correlation except for using Euclidean distances between sample elements rather than sample statistical moments (Székely and Rizzo 2013).

In order to determine the similarity measures by the SDC and MDC methods, one should initially obtain covariance distances as characteristic functions. For the SDC method, the sample covariance distance between E_u and E_d is given by:

$$D_{\text{cov}}^2(\mathbf{E_u}||\mathbf{E_d}) = \frac{1}{n^2} \sum_{i,j=1}^{n} A_{i,j} B_{i,j}, \tag{4.20}$$

in which

$$A_{i,j} = \left\| \mathbf{E_{u_i}} - \mathbf{E_{u_j}} \right\|_2 - \frac{1}{n} \sum_{k=1}^{n} \left\| \mathbf{E_{u_k}} - \mathbf{E_{u_j}} \right\|_2 - \frac{1}{n} \sum_{z=1}^{n} \left\| \mathbf{E_{u_i}} - \mathbf{E_{u_z}} \right\|_2$$

$$+ \frac{1}{n^2} \sum_{k,z=1}^{n} \left\| \mathbf{E_{u_k}} - \mathbf{E_{u_z}} \right\|_2, \tag{4.21}$$

$$B_{i,j} = \left\| \mathbf{E_{d_i}} - \mathbf{E_{d_j}} \right\|_2 - \frac{1}{n} \sum_{k=1}^{n} \left\| \mathbf{E_{d_k}} - \mathbf{E_{d_j}} \right\|_2 - \frac{1}{n} \sum_{z=1}^{n} \left\| \mathbf{E_{d_i}} - \mathbf{E_{d_z}} \right\|_2$$

$$+ \frac{1}{n^2} \sum_{k,z=1}^{n} \left\| \mathbf{E_{d_k}} - \mathbf{E_{d_z}} \right\|_2. \tag{4.22}$$

In these equations, $\|.\|_2$ refers to the L_2-norm or Euclidean norm. Similarly, one can determine the sample covariance matrices $D_{cov}{}^2(\mathbf{E_u},\mathbf{E_u})$ and $D_{cov}{}^2(\mathbf{E_d},\mathbf{E_d})$ based on Eq. (4.20). As a result, the SDC expression is presented as:

$$d_{SDC}(\mathbf{E_u}||\mathbf{E_d}) = \frac{D_{\text{cov}}^2(\mathbf{E_d},\mathbf{E_u})}{\sqrt{D_{\text{cov}}^2(\mathbf{E_u},\mathbf{E_u}) D_{\text{cov}}^2(\mathbf{E_d},\mathbf{E_d})}}. \tag{4.23}$$

The motivation for improving the SDC method and introducing the MDC approach is to deal with the correlation calculation of random samples when the dimensionality of matrices increases. The modification relies upon the development of sample covariance distance as follows:

$$D_{\text{cov}}^*(\mathbf{E_u}||\mathbf{E_d}) = \frac{1}{n(n-3)} \left[\sum_{i,j=1}^{n} A_{i,j}^* B_{i,j}^* - \frac{n}{n-2} \sum_{i=1}^{n} A_{i,i}^* B_{i,i}^* \right], \tag{4.24}$$

where

$$A_{i,j}^* = \begin{cases} \dfrac{n}{n-1}\left(A_{i,j} - \dfrac{\left\| \mathbf{E_{u_i}} - \mathbf{E_{u_j}} \right\|_2}{n} \right), & i \neq j \\[2ex] \dfrac{n}{n-1}\left(\dfrac{1}{n}\sum_{k=1}^{n} \left\| \mathbf{E_{u_k}} - \mathbf{E_{u_j}} \right\|_2 - \dfrac{1}{n^2}\sum_{k,z=1}^{n} \left\| \mathbf{E_{u_k}} - \mathbf{E_{u_z}} \right\|_2 \right), & i = j \end{cases} \tag{4.25}$$

and

$$B_{i,j}^* = \begin{cases} \dfrac{n}{n-1}\left(A_{i,j} - \dfrac{\left\| \mathbf{E}_{d_i} - \mathbf{E}_{d_j} \right\|_2}{n}\right), & i \neq j \\[3mm] \dfrac{n}{n-1}\left(\dfrac{1}{n}\sum_{k=1}^{n}\left\| \mathbf{E}_{d_k} - \mathbf{E}_{d_j} \right\|_2 - \dfrac{1}{n^2}\sum_{k,z=1}^{n}\left\| \mathbf{E}_{d_k} - \mathbf{E}_{d_z} \right\|_2\right), & i = j \end{cases} \tag{4.26}$$

Considering these expressions, the MDC method is formulated as:

$$d_{MDC}(\mathbf{E_u}\|\mathbf{E_d}) = \frac{D_{\mathrm{cov}}^*(\mathbf{E_d},\mathbf{E_u})}{\sqrt{D_{\mathrm{cov}}^*(\mathbf{E_u},\mathbf{E_u})D_{\mathrm{cov}}^*(\mathbf{E_d},\mathbf{E_d})}}. \tag{4.27}$$

Both d_{SDC} and d_{MDC} are scalar and take values in [0 1]. For damage detection, the SDC and MDC values close to zero are representative of damage occurrence since these indicate that there is no similarity between the residual matrices of the normal and damaged condition (Wang et al. 2013a).

4.6 Proposed Hybrid Distance Method

The new novelty detection methodology named as PKLD-MSD is a combination of Partition-based Kullback–Leibler divergence (PKLD) and MSD. This methodology is simultaneously intended to deal with the limitation of using randomly high-dimensional features in the distance computation by the PKLD method and make a decision on early damage detection by the MSD under varying the environmental and operational conditions.

4.6.1 Partition-Based Kullback–Leibler Divergence

The PKLD is a univariate distance measure that is intended to calculate dissimilarity between two random vectors as a universal divergence estimator (Wang et al. 2005). The key part of the PKLD method is to divide random samples into partitions or segments. Assum that the reference and target vectors $\mathbf{e_u}$ and $\mathbf{e_d}$ (e.g. the vectors of residuals of time series models related to the undamaged and damaged conditions) consist of n samples. Partitioning of the random variables in the PKLD method is based upon the maximum entropy approach as previously discussed in Sect. 4.4.1. Using the main outputs of this approach including the number of samples in each partition associated with the target vector (e.g. the residual vector of the damaged state $\mathbf{e_d}$) N_d, and the number of partitions or segments c, the equation of PKLD is expressed as:

$$d_{PKLD}(\mathbf{e_u}||\mathbf{e_d}) = \sum_{r=1}^{c-1} \frac{N_{u_r}}{n} \cdot \log\left(\frac{N_{u_r}}{N_d}\right) + \frac{N_{u_c}}{n} \cdot \log\left(\frac{N_{u_c}}{N_d + n\gamma_c}\right), \qquad (4.28)$$

where N_u is the number of sequences from $\mathbf{e_u}$ that falls into the domain of each partition of $\mathbf{e_d}$. To determine this amount, for example, assume that the rth partition of the arranged $\mathbf{e_d}$ is $0.05 < \mathbf{H_r} \le 0.5$ and 0.06, 0.45, 0.27, 0.11, 0.38, 0.09, 0.21, 0.43, 0.15, 0.08, 0.33 are the sequences of $\mathbf{e_u}$ in the domain of $\mathbf{H_r}$, in which case $N_u = 11$. It is important to point out that one does not need to arrange $\mathbf{e_u}$ in either ascending or descending order. Furthermore, $\gamma_c = (n\text{-}N_d c)/n$ is the correction factor for the last partition and N_{u_c} represents the number of samples in the last partition of $\mathbf{e_u}$.

4.6.2 PKLD-MSD

Although the PKLD method is an efficient statistical tool for measuring the similarity of two random high-dimensional data sets, it may fail in providing reliable damage detection results by features that are mainly affected by the operational and environmental variability. Furthermore, the MSD technique is applied to address this limitation in an unsupervised learning manner. For this purpose, one initially needs to learn a statistical model via the MSD technique by training data and then make a decision about the early damage detection using testing data. For the sake of convenience, Fig. 4.2 demonstrates the schematic representation of providing the training and testing datasets.

Assume that the structure in the training phase involves n_L normal or undamaged states designating $S_{N_1}, S_{N_2}, ...S_{N_L}$ with different operational and/or environmental conditions. This means that the structure in its normal behavior during the training or baseline phase is tested n_L times, each of which along with its vibration measurements and extracted DSFs is related to one undamaged state. For each normal state, the vibration time-domain responses are measured in n_T test measurements and n_S sensors. Furthermore, one supposes that $S_{C_1}, S_{C_2}, ...S_{C_U}$ denote n_U unknown or current structural conditions in the inspection phase with the same number of test measurements and sensors. The methodology for obtaining the training data is based on the computation of a distance matrix via PKLD between each normal condition with itself and the other undamaged states in the training phase by using their residual sets in all test measurements and all sensor locations. As an example, $\mathbf{D}_{Tr}^{S_{N_1}} \in \Re^{(n_S \times n_L n_T)}$ denotes the distance matrix of PKLD values for the state S_{N_1}, which is obtained by the distance calculations between S_{N_1} and $S_{N_1}...S_{N_L}$. The same distance matrix can be obtained from the other undamaged states. Once all PKLD matrices of all normal conditions in the training phase have been determined, the training data $\mathbf{D}_{Tr} \in \Re^{(n_S \times n_L n_L n_T)}$ is constructed by the combination of these matrices.

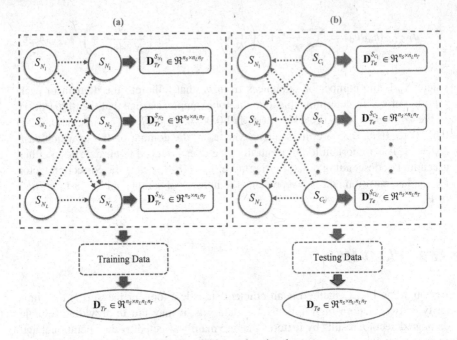

Fig. 4.2 The flowchart of providing the training and testing datasets

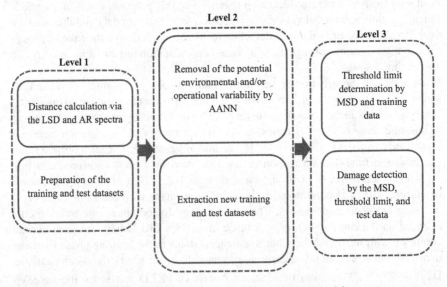

Fig. 4.3 The flowchart of the proposed multi-level distance-based methodology

In order to make the testing data in each unknown structural state of the inspection phase, the same procedure is repeated by computing the distance amounts between the current state and all the normal conditions in the baseline period by applying their residual sets at n_S sensors from n_T test measurements. For example, $\mathbf{D}_{Te}^{S_{C_1}} \in \Re^{(n_S \times n_L n_T)}$ refers to the PKLD matrix of the current state S_{C_1} in the inspection phase, which is made by calculating the distances between S_{C_1} and the structural conditions in the training phase $S_{N_1}...S_{N_L}$. Using all unknown structural conditions in the inspection stage, one can gain a distance matrix as $\mathbf{D}_{Te} \in \Re^{n_S \times n_{Te}}$ based on the combination of all PKLD matrices of these states, where $n_{Te} = n_U \times n_L \times n_T$.

The MSD is a powerful and well-known statistical distance metric that aims to measure the similarity between two multivariate datasets as described in Sect. 4.3.3. Considering the training dataset $\mathbf{D}_{Tr}$, one can learn an unsupervised learning model, which consists of the mean vector $\bar{\mathbf{d}}_{Tr} \in \Re^{n_S}$ and covariance matrix $\mathbf{\Sigma}_{Tr} \in \Re^{n_S \times n_S}$ of the training data. The MSD calculates the distance of each column vector of the testing data $\mathbf{d}_{Te} \in \Re^{n_S}$ from the trained model as follows:

$$D_{MSD}(l) = \left(\mathbf{d}_{Te} - \bar{\mathbf{d}}_{Tr}\right)^T \mathbf{\Sigma}_{Tr}^{-1} \left(\mathbf{d}_{Te} - \bar{\mathbf{d}}_{Tr}\right), \qquad l = 1, 2, ..., n_{Te}. \tag{4.29}$$

The fundamental principle behind the MSD method lies in the fact that if the feature vector is obtained from the damaged structure with the sources of operational and environmental variability, the vector will deviate from the mean of the normal condition implying the damage occurrence. On the contrary, if the feature vector comes from the undamaged status, even with the operational and environmental conditions, it will be close to the mean of the normal condition indicating that the structure is undamaged.

As mentioned above, the great advantage of the proposed PKLD-MSD method is to deal with the limitation of using random high-dimensional features such as the residuals of time series models. Such high-dimensional data samples are converted into low-dimensional spaces by determining the matrices $\mathbf{D}_{Tr}$ and $\mathbf{D}_{Te}$ through the PKLD. In fact, this distance measure serves as a dimensionality reduction technique. Moreover, the MSD undertakes the distance computation for damage detection by using the multivariate data sets $\mathbf{D}_{Tr}$ and $\mathbf{D}_{Te}$.

4.7 Proposed Multi-Level Distance Method

The proposed distance-based methodology contains the three main levels. The first level is based on the distance calculation between two spectra via the Log-Spectral Distance (LSD) and the providence of individual training and test datasets based on the structural states used in the baseline and monitoring phase. These datasets are then designated as the main multivariate features for damage detection. Having considered the multivariate training and test matrices, in the second level, an Auto-Associative Artificial Neural Network (AANN) algorithm is utilized to deal with or remove the effects of the potential environmental and/or operational variability conditions. Eventually, the third level is intended to apply the normalized training and test matrices for damage detection via the MSD technique. For the sake of convenience, Fig. 4.3 depicts the flowchart of the procedures of the feature extraction and multi-level distance-based methodology.

- **Level I: LSD**

The main objective of the first level of the proposed methodology is to provide multi-variate training and test datasets using the AR spectra estimated from the normal and current states. For this purpose, it is only necessary to calculate the (dis)similarity between those spectra by the LSD. This is asymmetric distance measure, which computes the discrepancy between two sets (vector) of frequency-domain data (Rabiner et al. 1993). Given the two spectra $P(\omega)$ and $\bar{P}(\omega)$, the LSD is given by:

$$LSD = \sqrt{\frac{1}{2\pi} \int_{-\pi}^{\pi} \left| \log \bar{P}(\omega) - \log P(\omega) \right|^2 d\omega} \qquad (4.30)$$

which can be rewritten as:

$$LSD = \sqrt{\frac{1}{2\pi} \int_{-\pi}^{\pi} \left| \log \frac{\bar{P}(\omega)}{P(\omega)} \right|^2 d\omega} \qquad (4.31)$$

If the spectra are in the discrete frequency domain, the LSD can be modified as follows (Deza and Deza 2014, p. 365):

$$D = \sqrt{\frac{1}{n_p} \sum_{i=1}^{n_p} \left| \log \bar{P}(i) - \log P(i) \right|^2} = \sqrt{\frac{1}{n_p} \sum_{i=1}^{n_p} \left| \log \frac{\bar{P}(i)}{P(i)} \right|^2} \qquad (4.32)$$

where np denotes the number of spectrum samples. The LSD becomes zero if and only if the spectra $P(\omega)$ and $\bar{P}(\omega)$ are exactly similar. On this basis, any difference between them leads to an LSD value larger than zero. Consider that $P(\omega)$ and $\bar{P}(\omega)$ are the AR spectra at a specific sensor location associated with the normal and current states of the structure. Any deviation of $\bar{P}(\omega)$ from $P(\omega)$ is most likely indicative of the occurrence of damage in that area.

This above-mentioned procedure is based on the distance calculation between two sets of spectra. In reality, there is more than one sensor location mounted on the structure. Moreover, the dynamic test process is usually repeated several times to have adequate data measurements. Assume that the structure of interest is deployed by n_s sensors and the dynamic test is repeated n_m times. In this case, one supposes that $S_1 \ldots S_c$ denote the n_c normal conditions available in the baseline phase. Therefore, the training datasets is $\mathbf{X} \in \mathbb{R}^{n_x \times n_s}$, where $n_x = n_m \times (n_c\text{-}1)$. Note that the main reason for using n_c-1 is that the distance calculation of the spectra in a normal condition with itself always corresponds to zero and one does not need to apply null values. Moreover, it worth remarking that each column of this matrix is the LSD value between the spectra $P(\omega)$ and $\bar{P}(\omega)$ for the two different structural normal conditions at the same sensor location.

Now, assume that S_u is the current structural condition in the monitoring phase, which is applied to inspect the status of the structure in terms of the damage occurrence. The process of distance calculation begins with computing the LSD values between the spectra of the normal conditions, $P(\omega)$, and the current state, $\bar{P}(\omega)$, as the same sensor location. This procedure continues to obtain all distance values for the n_c normal conditions at the n_s sensors and n_m test measurements. Hence, the test matrix is $\mathbf{Z} \in \mathbb{R}^{n_z \times n_s}$, where $n_z = n_m \times n_c$.

- **Level II: AANN**

The second level of the proposed distance-based methodology is to use an AANN algorithm for removing the potential environmental and/or operational variability conditions available in the distance amounts obtained from the first level. Once these amounts from the available sensors and all test measurements have been gathered in order to provide the multivariate training and test datasets, those are applied to an AANN algorithm. This algorithm is initially trained to learn the correlations between the data in the training set $\mathbf{X}$, which treat as an input. In this case, the network can quantify the unmeasured sources of the environmental and/or operational variability.

The AANN algorithm contains three hidden layers including mapping, bottleneck, and de-mapping. This algorithm is built up with a feed-forward neural network to perform the mapping and de-mapping, where the network outputs are simply the reproduction of the network inputs. Furthermore, the process of training the network is based on a Levenberg–Marquardt back-propagation technique. As long as the network has been trained, one can determine the network output called $\hat{\mathbf{N}}$. Finally, the residual between the input and output matrices are extracted as $\mathbf{E_x} = \mathbf{X} \text{-} \hat{\mathbf{N}}$ and used as the main features for damage detection.

During the monitoring period, it is not necessary to learn a new neural network. The trained network in the baseline phase using the training data is applied to remove the sources of the potential environmental and/or operational variability conditions from the test dataset $\mathbf{Z}$. In this regard, this dataset treats as an input applied to the trained network. The main objective is to determine the network output $\hat{\mathbf{Z}}$ and then extract the residual matrix $\mathbf{E_z} = \mathbf{Z} \text{-} \hat{\mathbf{Z}}$ used as the main features of the monitoring stage.

- **Level III: MSD**

The main goal of using the MSD in the third level of the proposed multi-level distance-based methodology is to make a decision about the status of the current state of the structure. For this purpose, one initially requires to train a statistical model from the training data, which is the residual $\mathbf{E_x}$ obtained from the AANN algorithm in the baseline phase. In this case, it is attempted to obtain the mean vector $\mathbf{m_x} \in \mathbb{R}^{n_s}$ and covariance matrix $\mathbf{S_x} \in \mathbb{R}^{n_s \times n_s}$ of $\mathbf{E_x}$. Subsequently, each feature vector of the test dataset $\mathbf{E_z}$ is applied to the MSD formulation in the following form:

$$d_M(i) = (\mathbf{e_z}(i) - \mathbf{m_x})^T \mathbf{S_x}^{-1}(\mathbf{e_z}(i) - \mathbf{m_x}), \qquad i = 1, 2, ..., n_z \qquad (4.33)$$

where $\mathbf{e_z}(i)$ is the ith vector of $\mathbf{E_z}$. In order to finalize the process of damage detection, it is necessary to compare each of the d_M values with a threshold limit. If the current state of the structure is undamaged, it is expected that the MSD quantity falls below the threshold; otherwise, one can infer that the structure suffers from damage and the current state is indicative of the damaged condition. To determine the threshold of interest, one only needs to apply each feature vector of $\mathbf{E_x}$ to Eq. (4.33) and obtain a vector of the d_M amounts regarding the normal conditions of the structure. Using the standard confidence interval under a significance level, the threshold limit is determined. For a 5% significance level, the threshold is based on the mean of the d_M quantities plus 1.96 multiplied by their standard deviation (Sarmadi and Karamodin 2020).

4.8   Threshold Estimation

Estimation of a threshold level for damage diagnosis is a crucial step of a novelty detection strategy. This level should be able to distinguish between the undamaged and damaged conditions to minimize the occurrence of Type I and Type II errors (Sarmadi and Karamodin 2020; Entezami and Shariatmadar 2017, 2019a, b; Entezami et al. 2018; Entezami et al. 2019a, b, 2020a, c). Type I error (false positive or false alarm) occurs when the structure is undamaged but the method of damage detection mistakenly alarms the occurrence of damage. On the contrary, Type II error (false negative or false detection) occurs when the structure suffers from damage but the method of interest incorrectly declares it is in its normal mode.

4.8.1   Standard Confidence Interval Limit

Generally, the threshold estimation in novelty detection is based on the outputs of the learned classifier by the training data. For this purpose, a probabilistic distribution is assumed for the training features to define the αth percentile as the threshold level. When the distribution of the features is normal or Gaussian, an efficient and practical approach is a statistical interval limit. This approach relies upon the use of some statistical properties of the normal probability distribution. Assume that $\mathbf{D} = [d_1 \dots d_{N_D}]$ is the vector of distance values obtained from the training data ($\mathbf{X}$) based on a statistical distance approach, in which case each distance quantity is representative of the similarity or discrepancy between the features of the undamaged condition(s). On this basis, the threshold τ is defined as the upper bound of the αth one-side confidence as follows:

$$\tau = \mu_D + Z_{\alpha/2} \frac{\sigma_D}{\sqrt{N_D}}, \tag{4.34}$$

where μ_D and σ_D denote the mean and standard deviation of the vector **D**. Moreover, $Z_{\alpha/2}$ refers to the confidence coefficient.

4.8.2 Monte Carlo Simulation

Monte Carlo simulation is a statistical method that aims to generate random samples (Kroese and Rubinstein 2012). This technique was initially proposed by Worden et al. (2000), who attempted to define the threshold limit for outlier analysis under multivariate feature datasets. The generalization of their approach is presented to establish a generalized Monte Carlo simulation approach to the threshold estimation.

The main idea of this approach to damage detection is to expand the outputs of classifiers obtained from the normal conditions of the structure by adding random noises. On this basis, a Gaussian white noise signal with zero mean and unit variance is separately added to the vibration measurements of the normal conditions. This process is repeated K times independently. Depending upon the statistical distance formulation used in novelty detection, the distance values are computed by using features extracted from the noisy vibration measurements of the normal conditions. In this regard, one can obtain a set of K distance values between the feature samples of normal states. Applying the standard confidence interval under the $\alpha\%$ significance level, one can compute a threshold value from the set of K distance values. The main advantage of the Monto Carlo simulation is its ability to determine the threshold limit for small samples. Nonetheless, it may suffer from computational inefficiency as its main disadvantage.

4.9 Conclusions

In this chapter of the book, some distance-based novelty detection methods have been proposed in order to establish the unsupervised learning strategy for feature classification under statistical decision-making. Initially, an overview of statistical distance has been mentioned to introduce the general problem of distance-based novelty detection and some conventional statistical distances. Four univariate distance methods namely KLDEPM, PAC, and RRC have been presented to use them for the process of damage localization based on the features extracted from the CBFE and RBFE approaches. For the early damage detection by using the high-dimensional features such as the residuals of time series models, the SDC, MDC, and PKLD-MSD methods have been presented.

References

Aggarwal CC (2016) Outlier analysis. Springer International Publishing

Baghiee N, Esfahani MR, Moslem K (2009) Studies on damage and FRP strengthening of reinforced concrete beams by vibration monitoring. Eng Struct 31(4):875–893

Barnett V, Lewis T (1994) Outliers in statistical data (probability & mathematical statistics). Wiley

Catbas FN, Gokce HB, Gul M (2012) Nonparametric analysis of structural health monitoring data for identification and localization of changes: Concept, lab, and real-life studies. Struct Health Monit:1–14

De Maesschalck R, Jouan-Rimbaud D, Massart DL (2000) The mahalanobis distance. Chemom Intell Lab Syst 50(1):1–18

Deraemaeker A, Worden K (2018) A comparison of linear approaches to filter out environmental effects in structural health monitoring. Mech Syst Sig Process 105:1–15

Deza MM, Deza E (2014) Encyclopedia of distances. Springer, Berlin Heidelberg

Dudley RM (1978) Central limit theorems for empirical measures. Ann Probab:899–929

Entezami A, Sarmadi H, Salar M, De Michele C, Arslan AN (2020a) A novel data-driven method for structural health monitoring under ambient vibration and high-dimensional features by robust multidimensional scaling. Struct Health Monit, In press

Entezami A, Shariatmadar H (2017) An unsupervised learning approach by novel damage indices in structural health monitoring for damage localization and quantification. Struct Health Monit 17(2):325–345

Entezami A, Shariatmadar H (2019) Damage localization under ambient excitations and non-stationary vibration signals by a new hybrid algorithm for feature extraction and multivariate distance correlation methods. Struct Health Monit 18(2):347–375

Entezami A, Shariatmadar H (2019) Structural health monitoring by a new hybrid feature extraction and dynamic time warping methods under ambient vibration and non-stationary signals. Measurement 134:548–568

Entezami A, Shariatmadar H, Karamodin A (2018) Data-driven damage diagnosis under environmental and operational variability by novel statistical pattern recognition methods. Struct Health Monit 18(5–6):1416–1443

Entezami A, Shariatmadar H, Mariani S (2019a) Low-order feature extraction technique and unsupervised learning for SHM under high-dimensional data. In: MORTech 2019, 5th international workshop on reduced basis, POD and PGD model reduction techniques. FRA, pp 72–73

Entezami A, Shariatmadar H, Mariani S (2019b) A novelty detection method for large-scale structures under varying environmental conditions. In: Sixteenth International Conference on Civil, Structural & Environmental Engineering Computing (CIVIL-COMP2019), 16–19 September, 2019, Riva del Garda, Italy

Entezami A, Shariatmadar H, Mariani S (2020) Early damage assessment in large-scale structures by innovative statistical pattern recognition methods based on time series modeling and novelty detection. Adv Eng Softw 150:102923

Entezami A, Shariatmadar H, Mariani S (2020) Fast unsupervised learning methods for structural health monitoring with large vibration data from dense sensor networks. Struct Health Monit 19(6):1685–1710

Entezami A, Shariatmadar H, Sarmadi H (2020) Condition assessment of civil structures for structural health monitoring using supervised learning classification methods. Iranian J Sci Technol, Trans Civ Eng 44(1):51–66

Farrar CR, Worden K (2013) Structural health monitoring: a machine learning perspective. Wiley

Figueiredo E, Cross E (2013) Linear approaches to modeling nonlinearities in long-term monitoring of bridges. J Civ Struct Health Monit 3(3):187–194

Kroese DP, Rubinstein RY (2012) Monte carlo methods. Wiley Interdiscip Rev: Comput Stat 4(1):48–58

Kullback S, Leibler RA (1951) On information and sufficiency. Ann Math Stat 22(1):79–86

Markou M, Singh S (2003) Novelty detection: a review—Part 1: statistical approaches. Sig Process 83(12):2481–2497

Mustapha S, Lu Y, Li J, Ye L (2014) Damage detection in rebar-reinforced concrete beams based on time reversal of guided waves. Struct Health Monit 13(4):347–358

Nguyen T, Chan T, Thambiratnam D (2014) Controlled Monte Carlo data generation for statistical damage identification employing Mahalanobis squared distance. Struct Health Monit 13(4):461–472

Nguyen T, Chan TH, Thambiratnam DP (2014) Field validation of controlled Monte Carlo data generation for statistical damage identification employing Mahalanobis squared distance. Struct Health Monit 13(4):473–488

Nowak AS, Collins KR (2012) Reliability of structures, 2nd edn. Taylor & Francis

Pastor M, Binda M, Harčarik T (2012) Modal assurance criterion. Procedia Eng 48:543–548

Pimentel MAF, Clifton DA, Clifton L, Tarassenko L (2014) A review of novelty detection. Sig Process 99:215–249

Rabiner L, Rabiner LR, Juang BH (1993) Fundamentals of Speech Recognition. PTR Prentice Hall

Sarmadi H, Entezami A (2020) Application of supervised learning to validation of damage detection. Archive of Applied Mechanics

Sarmadi H, Entezami A, Daneshvar Khorram M (2020) Energy-based damage localization under ambient vibration and non-stationary signals by ensemble empirical mode decomposition and Mahalanobis-squared distance. J Vib Control 26(11–12):1012–1027. https://doi.org/10.1177/1077546319891306

Sarmadi H, Entezami A, Saeedi Razavi B, Yuen K-V (2020b) Ensemble learning-based structural health monitoring by Mahalanobis distance metrics. Struct Control Health Monit:e2663

Sarmadi H, Karamodin A (2020) A novel anomaly detection method based on adaptive Mahalanobis-squared distance and one-class kNN rule for structural health monitoring under environmental effects. Mech Syst Sig Process 140:106495

Székely GJ, Rizzo ML (2013) The distance correlation t-test of independence in high dimension. J Multivar Anal 117:193–213

van der Vaart AW (2000) Asymptotic Statistics. Cambridge University Press

Wang D, Song H, Zhu H (2013) Numerical and experimental studies on damage detection of a concrete beam based on PZT admittances and correlation coefficient. Constr Build Mater 49:564–574

Wang D, Ye L, Su Z, Lu Y, Li F, Meng G (2009) Probabilistic damage identification based on correlation analysis using guided wave signals in aluminum plates. Struct Health Monit 9(2):133–144

Wang J, Tsang WW, Marsaglia G (2003) Evaluating Kolmogorov's distribution. J Stat Softw 8(18):1–4

Wang Q, Kulkarni SR, Verdú S (2005) Divergence estimation of continuous distributions based on data-dependent partitions. IEEE Trans Inf Theory 51(9):3064–3074

Wang X, Mueen A, Ding H, Trajcevski G, Scheuermann P, Keogh E (2013) Experimental comparison of representation methods and distance measures for time series data. Data Min Knowl Disc 26(2):275–309

Worden K, Manson G, Fieller NRJ (2000) Damage detection using outlier analysis. J Sound Vib 229(3):647–667

Yeager M, Gregory B, Key C, Todd M (2019) On using robust Mahalanobis distance estimations for feature discrimination in a damage detection scenario. Struct Health Monit 18(1):245–253

Chapter 5
Applications and Results

5.1 Introduction

Due to the importance and necessity of SHM in civil structures and dealing with some major challenging issues, several methods were proposed in the previous chapters. These methods can be divided into two parts including the feature extraction from stationary and non-stationary vibration signals and statistical decision-making for damage diagnosis. The great advantage of these methods is that no finite element model and structural model updating are needed. This is because those are directly data-driven algorithms and it is not necessary to have any information about the numerical model of the structure or its details. Using the only measured vibration data, it is feasible to extract damage-sensitive features from the proposed feature extraction methods and diagnose damage in terms of damage detection, localization, and quantification. In this chapter of the book, the proposed methods are validated by some benchmark structures including the four-story laboratory frame of the LANL, the second phase of the IASC-ASCE structure, the wooden truss bridge, and the cable-stayed bridge. The experimental validations and performance evaluations are carried out by verifying some proposed methods via one of the benchmark structures. The proceeding parts of this chapter of the book are as follows: the proposed DFBFE, PAC, and RRC methods are validated by the LANL frame in Sect. 5.2. In Sects. 5.3 and 5.4, the IASC-ASCE structure is considered to verify the proposed FRBFE, ICEEMDAN-ARMA, KLDEPM, SDC, and MDC approaches. Section 5.5 is intended to validate the proposed automatic model identification via the ARMAsel algorithm and the hybrid novelty detection method PKLD-MSD by the cable-stayed bridge. The wooden truss bridge is applied to verify the proposed spectral-based and multi-level distance-based methods in Sect. 5.6. Eventually, Sect. 5.7 summarizes the conclusions of this chapter.

5.2 Validation of the DRBFE, PAC, and RRC Methods by the LANL Laboratory Frame

The three-story laboratory frame was comprised of aluminum columns (height 177 mm, width 25 mm, and thickness 6 mm) and aluminum plates (length 305 mm × 305 mm, and thickness 25 mm) as shown in Fig. 5.1. At each story, four columns were connected to the top and bottom of the plate assembled using bolted joints. Four accelerometers (sensors 2–5) were mounted on the floors, as can be observed in Fig. 5.1, to measure acceleration time histories. A random vibration load was applied by an electrodynamic shaker to the base floor along the centerline of the frame, to excite the frame. The acceleration time histories were sampled at 320 Hz for 25.6 s in duration discretized into 8192 data points and 0.003125 s time interval.

To apply damage to the frame, a center column (height 150 mm and cross-section 25 mm × 25 mm) was suspended from the third floor. This column was connected with a bumper mounted on the second floor, in which the gap in the bumper was adjustable for defining various damage cases. The source of damage was a simulation of breathing cracks to make nonlinear behavior. The acceleration time histories were measured under 17 structural state conditions as described in Table 5.1.

These conditions were categorized into five groups. The first group referred to an ideal undamaged or baseline condition, in which there were no adverse or deceptive changes in the frame (the state 1). The second and third groups represented different normal conditions under diverse EOV (the states 2–3 and 4–9). The fourth group included five damaged conditions with different gap distances (the states 10–14). Eventually, the last group referred to three damaged states by considering operational variability conditions (the states 15–17). Note that states 10, 15, and 16 present the lowest levels of damage severities due to the same large gaps, while the state 14

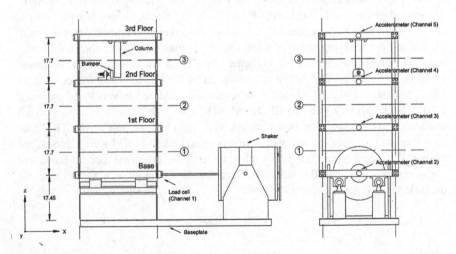

Fig. 5.1 The shaker and sensor locations

Table 5.1 The structural state conditions in the laboratory frame (Figueiredo et al. 2009)

State	Condition	Description
1	Undamaged	Baseline
2	Undamaged	Added mass of 1.2 kg at the base
3	Undamaged	Added mass of 1.2 kg at the 1st floor
4	Undamaged	87.5% stiffness reduction in one column of the 1st inter-story
5	Undamaged	87.5% stiffness reduction in two columns of the 1st inter-story
6	Undamaged	87.5% stiffness reduction in one column of the 2nd inter-story
7	Undamaged	87.5% stiffness reduction in two columns of the 2nd inter-story
8	Undamaged	87.5% stiffness reduction in one column of the 3rd inter-story
9	Undamaged	87.5% stiffness reduction in two columns of the 3rd inter-story
10	Damaged	Distance between bumper and column tip 0.20 mm
11	Damaged	Distance between bumper and column tip 0.15 mm
12	Damaged	Distance between bumper and column tip 0.13 mm
13	Damaged	Distance between bumper and column tip 0.10 mm
14	Damaged	Distance between bumper and column tip 0.05 mm
15	Damaged	Bumper 0.20 mm from column tip, 1.2 kg added at the base
16	Damaged	Bumper 0.20 mm from column tip, 1.2 kg added at the 1st floor
17	Damaged	Bumper 0.10 mm from column tip, 1.2 kg added at the 1st floor

has the highest one owing to the smallest gap. This section focuses on using the information and measured vibration responses of the LANL laboratory frame in order to demonstrate the effectiveness and performance of the following proposed methods:

- Box-Jenkins methodology for model identification (Sect. 2.5.1)
- Robust order selection by an iterative algorithm (Sect. 2.6.1)
- Robust and optimal order selection by a two-stage iterative algorithm (Sect. 2.6.2)
- Feature extraction by the conventional RBFE and CBFE (Sect. 2.8)
- Feature extraction by the proposed DRBFE (Sect. 2.9.1)
- Distance-based damage indices PAC and RRC (Sects. 4.4.2 and 4.4.3)

5.2.1   Model Identification by Box-Jenkins Methodology

Before identifying an appropriate model, it is important to assess the nature of vibration responses. One efficient approach is to use statistical hypothesis tests. In this section, *Leybourne-McCabe* (LMC) test is applied to evaluate the stationarity of the acceleration responses of the LANL frame. This hypothesis was proposed by Leybourne and McCabe (1994) for analyzing the stationarity or non-stationarity of univariate time series samples. In essence, this test assesses the null hypothesis that

Table 5.2 The p-values of the LMC test at all sensors of the states 1 and 14

State no.	Sensor no.			
	2	3	4	5
1	0.1	0.1	0.1	0.1
14	0.1	0.1	0.1	0.1

the time series data is a trend stationary AR process against the alternative hypothesis that it is a non-stationary ARIMA process. The outputs of the LMC test include the logical values $\mathbb{H}_0$ (the null hypothesis) or $\mathbb{H}_1$ (the alternative hypothesis), p-value, the test statistic (Q_{LMC}), and c-value. The null hypothesis of the LMC test refers to the stationarity of the time series. Under this hypothesis and a predefined significance level (α), Q_{LMC} is smaller than the c-value and the test p-value is larger than α. Considering the 5% significance level ($\alpha = 0.05$), Table 5.2 lists the p-values of the LMC test at all sensors of the states 1 and 14. It is observed that the vibration responses at all sensors of the states 1 and 14 are stationary since all p-values are larger than 0.05. Hence, it is feasible to use the time-invariant linear models for response modeling and feature extraction.

To identify the most appropriate model, Fig. 5.2 illustrates the plots of ACF and PACF concerning the acceleration time histories at the sensor 5 of the states 1 and 14. As can be seen, the ACFs (the left plots) in these conditions have exponentially decreasing forms and do not tend to decay zero, whereas the PACFs (the right plots) in the baseline and damaged conditions gradually become zero after approximately the 30th lag. According to the Box-Jenkins methodology, these results confirm that the AR model is proper for acceleration time histories. It should be noted that this conclusion is also valid for the acceleration responses at the other sensors in the other structural states.

5.2.2 AR Order Selection by the Proposed Iterative Algorithm

Based on the proposed iterative order selection technique in Sect. 2.6.1, the robust order of each vibration signal at each sensor is determined by analyzing the model residuals through the LBQ test. The p-values of AR residuals and the robust orders at all sensors of the laboratory frame in the baseline condition are shown in Fig. 5.3, where the dashed red lines refer to the significance value $\alpha = 0.05$. It is discerned that the robust orders, which enable the AR models to generate uncorrelated residuals, are equivalent to iterations that their p-values are larger than 0.05. On this basis, the robust orders at the sensors 2–5 are identical to 50, 51, 38, and 41, respectively. These quantities confirm that the residuals of AR models are uncorrelated since there is no enough numerical evidence (p-value < 0.05) for the rejection of the null hypothesis. Since the first p samples of the residual vector of AR(p) are always zero, it is possible to eliminate them from the process of robust order selection. On this

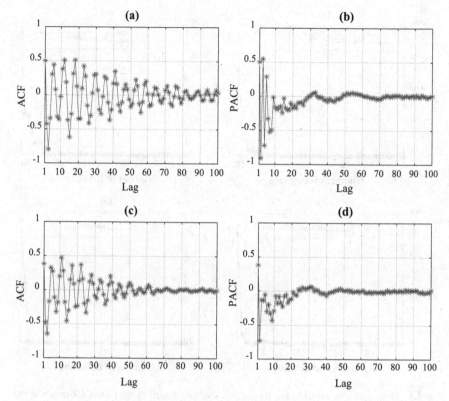

Fig. 5.2 Model identification by Box-Jenkins methodology at Sensor 5: **a** ACF in the state 1, **b** PACF in the state 1, **c** ACF in the state 14, **d** PACF in the state 14

basis, the above-mentioned orders are gained by the elimination of the zero quantities of residual vector.

Another evaluation for making sure of the accuracy and adequacy of the AR models obtained from the robust orders is to examine the correlation of residuals. To implement this procedure and also provide a comparative analysis, Fig. 5.4 shows the plots of ACF associated with the residuals of AR(5), AR(15), AR(30), and AR(41). It is important to point out that the first three AR orders were previously chosen by Figueiredo et al. (2010) to evaluate their influences on damage detection. As can be observed in Fig. 5.4, there are strongly correlated residuals in the AR(5) since the samples of ACF in this model extremely exceed the upper and lower confidence bounds. Although increasing the orders of the AR model from 15 to 30 reduces the correlation among the residuals, the samples of ACF in these models still exceed the confidence limits. However, the ACF plot of the residuals of AR(41) indicates that the robust order potentially enables the model to extract uncorrelated residuals since all ACF samples are within the lower and upper bounds. Therefore, the AR models obtained from the robust orders are accurate and adequate to use in the step of feature extraction.

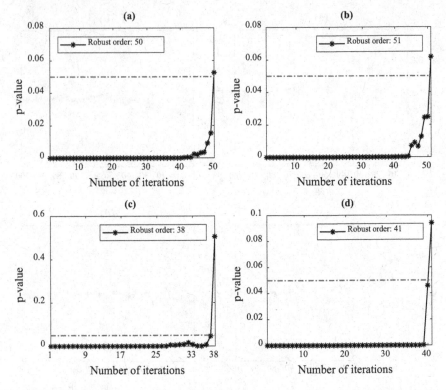

Fig. 5.3 The *p*-values of the LBQ test for the AR residuals gained by the robust AR orders in the baseline condition: **a** Sensor 2, **b** Sensor 3, **c** Sensor 4, **d** Sensor 5

5.2.3 AR Order Selection by the Proposed Two-Stage Iterative Algorithm

In the previous section, the robust AR orders have been chosen by using the LBQ test. As discussed in Sect. 2.6.2, the proposed two-stage iterative algorithm with the aids of the LBQ and Durbin-Watson hypothesis tests is used to select the robust and optimal orders for the acceleration responses of the sensors 2–5 of the laboratory frame. The results are presented in Table 5.3.

As can be observed from Table 5.3, the *p*-values in the first iterative algorithm are greater than 0.05, which means that the maximum orders of the AR models are exactly able to make the uncorrelated residuals. Moreover, one can perceive that all Durbin-Watson statistic values are larger than 2. Such observations prove that both the maximum and optimal orders lead to the uncorrelated residuals. It is significant to mention that the maximum orders based on the LBQ test have been determined without the removal of zero values of the residual vector in each iteration. For this reason, the maximum orders of the AR models at the sensors 2–5 obtained from the

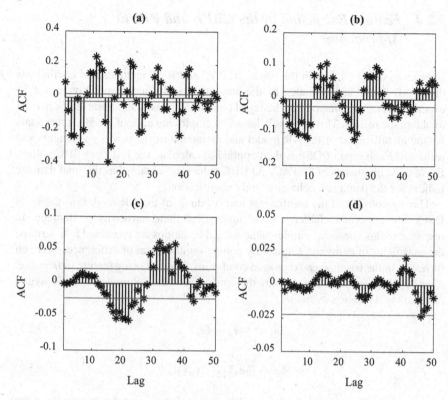

Fig. 5.4 The comparison of the correlation of residuals by the ACF at the sensor 5 in the baseline condition: **a** AR(5), **b** AR(15), **c** AR(30), **d** AR(41)

Table 5.3 The maximum and optimal orders of AR models in the baseline condition

Sensors	Maximum orders	p-values	Optimal orders	DW
2	60	0.5337	26	2.6871
3	54	0.3063	22	3.1129
4	46	0.0722	17	2.1088
5	58	0.1903	24	3.8009

LBQ test in the first step of the two-stage iterative algorithm are slightly different from the same orders gained by the LBQ test.

5.2.4 Feature Extraction by the CBFE and RBFE Approaches

In order to use the DSFs in the proposed PAC and RRC methods, the coefficients of the AR model are estimated by the non-recursive LS technique. Notice that 50 experiments of undamaged and damaged conditions in each structural state are used to obtain the results. This process is based on employing each of the 50 experiments for the identification of the AR model and extraction of the model coefficients and residuals. Each type of DSF is then applied to calculate the PAC and RRC values. Eventually, the means of 50 PAC and RRC values are obtained as the final damage indices for the damage localization and quantification.

Having considered the coefficients and residuals of the AR models as the main DSFs, the important challenge is to understand their sensitivity to damage. In response to this question, simple mathematical equations are presented to determine the sensitivity of extracted features by computing the norm of differences between the DSFs in the baseline and damaged conditions. The following feature indices (i.e. δ_c and δ_r) are formulated to measure the sensitivity of the AR model coefficients and residuals to damage:

$$\delta_1 = \|\mathbf{\Theta_u} - \mathbf{\Theta_d}\|_2, \qquad (5.1)$$

$$\delta_\nabla = |\|\mathbf{e_u}\|_2 - \|\mathbf{e_d}\|_2|. \qquad (5.2)$$

At each sensor location, both δ_c and δ_r give scalar amounts indicating the norm of the difference between the model coefficients and residuals in the undamaged and damaged conditions. Figure 5.5 shows the amounts of these indices at the sensors 2–5 in the states 10–14.

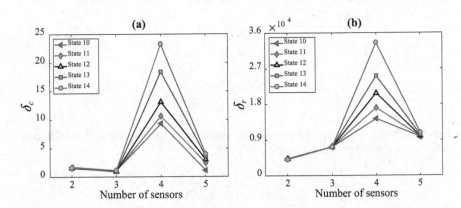

Fig. 5.5 Evaluating the sensitivity of extracted DSFs in the states 10–14: **a** the AR coefficients obtained from the CBFE approach, **b** the AR residuals obtained from the RBFE approach

The observations in Fig. 5.5a, b indicate that there are obvious differences between the feature indices at the sensor 4, where is the location of damage due to exciting the bumper (the source of damage) at the vicinity of this sensor. At this location, the quantities of δ_c and δ_r increase with increasing the level of damage from the state 10 (the lowest level of damage severity) to the state 14 (the highest level of damage severity). An important note is that the amounts of these indices are approximately invariant at the sensors 2 and 3. In Fig. 5.5a, there are a few variations at the location of the sensor 5, whereas the corresponding observations at this sensor in Fig. 5.5b indicate no alterations. Such results prove that the coefficients and residuals of the AR models extracted from the CBFE and RBFE approaches with the aid of the proposed two-stage iterative order selection method are sensitive to damage.

5.2.5   Feature Extraction by the Proposed DRBFE Approach

The main benefit of the proposed DRBFE method is the ability to simultaneously select the adequate order and extract the model residuals as the DSFs. According to this approach, the initial orders of the state 1 are 36, 28, 12, and 16 regarding the sensors 2–5, respectively. These orders are obtained from the well-known BIC technique. In this regard, the improved orders at these sensors at the same state include 46, 40, 31, and 35, respectively. With these numbers, the maximum order is 46; hence, AR(46) is fitted to the acceleration responses acquired from all sensors in the selected undamaged and damaged conditions for the extraction of the model residuals as the DSFs. A comparative analysis is also carried out to evaluate the correlation of the residual sequences obtained from the initial and improved AR orders. For this comparison, Table 5.4 lists the p-values and the statistics of the LBQ test using the residuals sequences of the sensors 3 and 5 gained by the initial and improved orders. The amounts of the test statistics and p-values are based on the 5% significance level, in which case the c-value corresponds to 31.4104.

As can be seen, the test statistics and p-values at the sensors 3 and 5 related to the initial order are larger than the c-value and smaller than the significance level, respectively. These consequences prove that the state-of-the-art BIC technique fails in extracting the uncorrelated residuals. On the contrary, one can discern that the test statistics and p-values concerning the improved order are smaller than the c-value and larger than 0.05, in which case one can make sure of generating uncorrelated residuals via the improved order.

Table 5.4 Comparison of the performances of the initial and improved orders in generating the uncorrelated residuals by the statistic and p-value of the LBQ test under the 5% significance level

Order level	Sensor no.	Test statistic	p-Value
Initial	3	44.8185	0.0175
	5	33.0807	0.0069
Improved	3	29.1250	0.1730
	5	27.7470	0.2805

5.2.6 Damage Diagnosis by PAC and RRC

In order to have a more reliable damage localization procedure, it is very appropriate to determine a threshold level. For this purpose, the DSFs of the states 1–9 in the training phase obtained from the conventional CBFE and RBFE techniques are used in the PAC and RRC methods. Applying the 95% confidence interval of the PAC and RRC values, the threshold limits based on Eq. (4.34) for the PAC and RRC methods are 0.6086 and 0.3715, respectively. For the identification of damage location, the coefficients and residuals of AR models in the state 1 are employed as the features of undamaged state (Θ_u and e_u) in both the PAC and RRC methods. In contrast, Θ_d and e_d of each of the states 10–17 are utilized as the characteristics of the damaged conditions.

Figure 5.6a illustrates the results of damage localization by the proposed PAC method in the damaged conditions. As can be observed, all PAC values at the sensor 4 are less than the threshold limit, which implies that this area of the laboratory frame is the location of the damage. This is because the bumper (the source of damage) was placed in the vicinity of the sensor 4. In this figure, the PAC quantities at the only sensor 4 are close to zero, whereas the other PAC values are far from the threshold limit and approximately near to one indicating the undamaged locations. In Fig. 5.6b, the amounts of RRC at the sensor 4 exceed the threshold limit, which means that this area of the frame is damaged. In addition, the RRC values at the other sensors are less than the threshold level indicating the undamaged areas in the frame. Eventually, the results of damage localization lead to the conclusion that the proposed PAC and

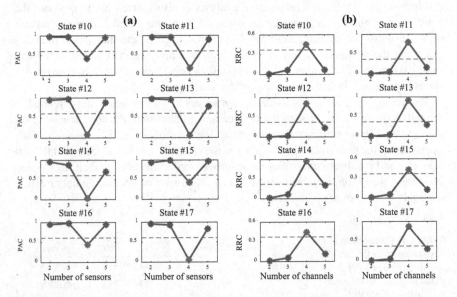

Fig. 5.6 Damage localization in the laboratory frame: **a** PAC, **b** RRC

Fig. 5.7 The four-story steel structure of the IASC-ASCE SHM benchmark problem (Dyke et al. 2003)

RRC are capable of identifying the location of damage even in the presence of the EOV conditions (Figs. 5.7 and 5.8).

5.3 Validation of the FRBFE and KLDEPM Methods by the IASC-ASCE Structure Under the Shaker Excitation

In this section, the experimental datasets of the IASC-ASCE benchmark structure in its second phase (Dyke et al. 2003) are utilized to verify the robustness and reliability of some proposed methods. The structure was located in the Earthquake Engineering

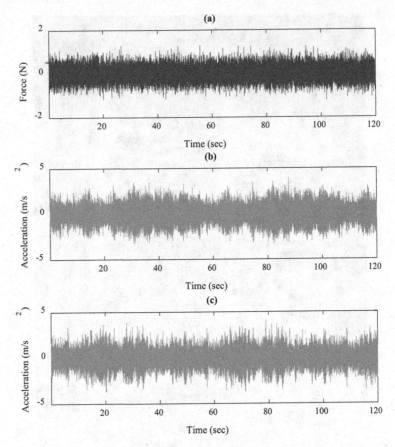

Fig. 5.8 **a** The force excitation signal generated by the shaker, **b** the acceleration response of the sensor 15 in the healthy state, **c** the acceleration response of the sensor 15 in the damaged state

Research Laboratory at the University of British Columbia in Canada. It was a four-story steel structure in a scale-model as a 2-bay-by-2-bay steel frame with 2.5 × 2.5 m in plan and 3.6 m in tall as shown in Fig. 5.9. For dynamic tests, the structure was mounted on a concrete slab outside of the structural testing laboratory to simulate typical ambient vibration conditions. The members were hot-rolled grade 300 W steel with the nominal yield stress 300 MPa. The columns and floor beams were constructed by B100 × 9 and S75 × 11 sections, respectively. The nine columns were bolted to a steel base frame and the lower flanges of two of the base beams were encased in concrete, which fixed the steel frame to the concrete slab. In each bay, the bracing system consisted of two 12.7 mm diameter threaded steel rods placed in parallel along the diagonal. To make the mass distribution reasonably realistic, the floor of each bay was placed by one slab. On this basis, four slabs with the mass of 1000 kg were placed on the first, second, and third floors and four 750 kg slabs on the fourth floor.

Fig. 5.9 Determination of the AR order in all five measurements of Case 1: **a** the improved algorithm in the proposed FRBFE method, **b** the original technique presented in Sect. 2.6.1, **c** BIC

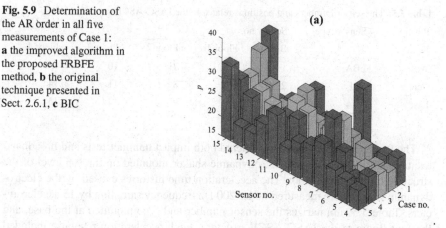

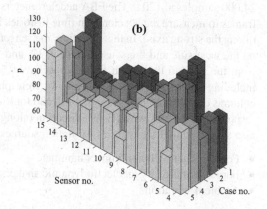

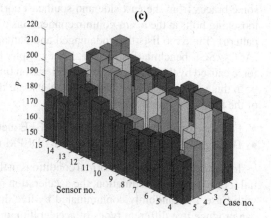

Table 5.5 The sensor numbers and positions related to the IASC-ASCE structure

Position	Sensor type	Sensor no.				
		Base	Floor 1	Floor 2	Floor 3	Floor 4
West	FBA	1	4	7	10	13
Center	EPI	2	5	8	11	14
East	FBA	3	6	9	12	15

The force excitations were considered both impact hammer tests and broadband excitations provided by an electrodynamic shaker mounted on the top level of the structure on the fourth floor. The acceleration time histories caused by the electro-dynamic shaker were measured under 200 Hz frequency sampling by 15 accelerometers. Table 5.5 summarizes the sensor number and type mounted at the base, and the four floors of the IASC-ASCE structure. Each acceleration response included 24,000 samples at 120 s. The FBA accelerometers were located on the east and west frames to measure the acceleration time histories in the north-south (N–S) direction (along the strong axis). In this direction, the sensors 1, 4, 7, 10, and 13 were mounted on the west side and the sensors 3, 6, 9, 12, and 15 were installed on the east face from the base to the fourth floor, respectively. Moreover, the EPI accelerometers including the sensors 2, 5, 8, 11, and 14 were placed in the vicinity of the central columns of the base, first, second, third, and fourth floors to measure the acceleration responses in the east-west (E–W) direction (along the weak axis). The structure was also subjected to three types of excitation sources including:

- Force excitation by an impact hammer.
- Force excitation by an electrodynamic shaker.
- Ambient vibration.

A total nine damage scenarios were simulated on the structure including removing some braces from the east side and southeast corner (the first damage pattern) and loosening bolts at the beam-column connections of the east side (the second damage pattern). Table 5.6 lists the undamaged and damaged cases of these patterns of the IASC-ASCE benchmark structure. For example, Fig. 5.10 illustrates the excitation force caused by the shaker and two acceleration time-domain responses at the sensor 9. In this section, it is attempted to demonstrate the effectiveness and performance of the following proposed methods:

- Feature extraction by the proposed FRBFE method (Sect. 2.9.2).
- Damage diagnosis by the proposed KLDEPM method (Sect. 4.4.1).

In order to have adequate normal conditions in the training phase, which are essential for the threshold estimation, the acceleration response of each sensor associated with Case 1 is randomly contaminated by five different noise levels. As such, one can produce five different types of acceleration datasets for the undamaged state of the IASC-ASCE structure, which are equivalent to five normal conditions or five test measurements of Case 1 in the training stage.

Table 5.6 The undamaged and damaged cases of the IASC-ASCE structure in the second phase

Pattern no.	Case no.	Label	Description
1	1	Healthy	Fully braced configuration—no damage
	2	Damaged	Removing all braces from the floors 1–4 on the east side
	3	Damaged	Removing all braces from the floors 1–4 on the southeast corner
	4	Damaged	Removing the braces from the floors 1 and 4 on the southeast corner
	5	Damaged	Removing the braces from the floor 1 on the southeast corner
	6	Damaged	Removing the braces from all floors on the east side and the floor 2 on the north face
2	7	Healthy	Removing all braces from the structure
	8	Damaged	Case 7 + loosened bolts on the floors 1–4 at both ends of the beam on the east face and north side
	9	Damaged	Case 7 + loosened bolts on the floors 1 and 2 on the east face and north side

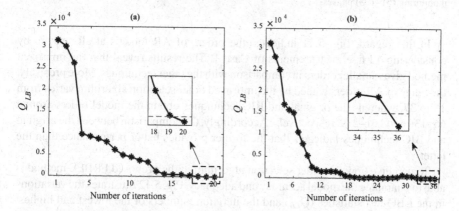

Fig. 5.10 The LBQ test statistics in the first measurement of Case 1: **a** Sensor 9, **b** Sensor 15

5.3.1 Feature Extraction by FRBFE

Based on the first stage of the proposed FRBFE method, it is necessary to select an adequate order of the AR model and simultaneously extract the model residuals at each sensor as the main DSFs of the undamaged state in the training phase. Concerning the iterative algorithm of the FRBFE method, one can state that the process of order selection in the first stage is, in fact, an improvement on the original iterative order determination approach as discussed in Sect. 2.6.1. Therefore, it is appropriate to compare the original and improved order selection approaches. Both iterative algorithms are also compared with the state-of-the-art BIC technique.

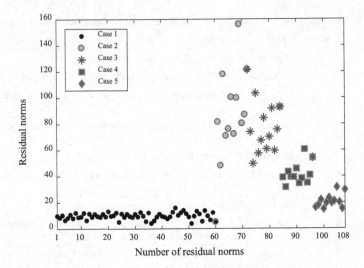

Fig. 5.11 Variations in the residual L_2-norms of the structural conditions in the training (1–60) and monitoring (61–108) phases

In this regard, Fig. 5.11 indicates the orders of AR models at all sensors by considering all five measurements of Case 1. The results reveal that the improved method gives smaller orders in comparison with the other techniques. More precisely, the range of AR orders gained by the improved order selection algorithm varies from 18 to 37, whereas the original and BIC techniques obtain the model orders around 65–130 and 161–219, respectively. Accordingly, the comparison between the original and BIC techniques indicates that the former provides better performance than the latter.

To perceive how the order selection algorithm in the proposed FRBFE method is able to guarantee the model accuracy and adequacy, Fig. 5.12 illustrates the variations in the LBQ test statistics (Q_{LB}) and the iteration numbers of the lowest and highest AR orders of the sensors 9 and 15 in the first test measurements of Case 1. As can be seen, the Q_{LB} values of the iteration numbers 20 and 36 are smaller than the test c-value, which corresponds to 31.4104 by adopting the 5% significance limit. The same conclusion is achievable for the other sensors and measurements. For the other comparative analyses, Tables 5.7 and 5.8 present the computational time of AR order selection and a complete process of RBFE. In Table 5.7, the vibration responses from all sensors in the five measurements of Case 1 are considered to compute the time needed for the determination of AR orders. The vibration responses of Case 2 and the only first measurement of Case 1 are utilized in Table 5.8 to measure the computational time of the residual extraction in the training and monitoring phases by using the proposed FRBFE and conventional RBFE methods.

It is seen in Table 5.7 that the improved order selection algorithm in the proposed FRBFE method requires very little computational time compared to the other techniques. The comparison between the original and BIC techniques indicates that the

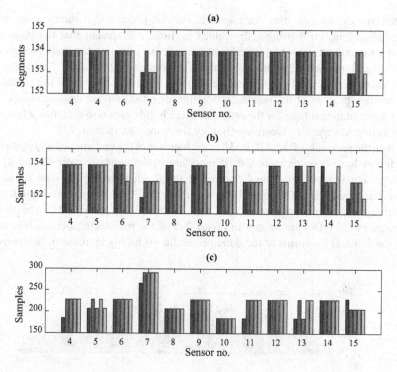

Fig. 5.12 Segmentation of the residual vectors of Cases 2–5 at all sensors: **a** number of segments, **b** number of samples of $\mathbf{e}_{d_z}$ within the first c-1 segments, **c** number of samples of $\mathbf{e}_{d_z}$ within the last segment

Table 5.7 The computational time for AR order selection in the IASC-ASCE structure obtained from the improved algorithm in the proposed FRBFE method, the original technique presented in Sect. 2.6.1, and BIC

Case	Improved	Original	BIC
1	5.84	97.94	496.80
2	5.72	97.27	493.43
3	5.78	96.86	492.34
4	5.69	97.27	497.39
5	5.71	97.58	555.95

Table 5.8 The computational time for a complete process of RBFE in the IASC-ASCE structure (proposed RBFE refers to the proposed FRBFE method)

Process	Improved	Original	BIC
Order selection in the training phase	5.85	97.94	496.80
Residual extraction in the training phase	0.00	438.28	621.09
Residual extraction in the monitoring phase	354.20	202.46	256.89

latter takes the longest time for choosing the AR order and it seems to be very time-consuming. Furthermore, the values in Table 5.8 present that the proposed FRBFE method provides better performance in terms of computational time against the conventional RBFE approach. Although the step of *"Residual Extraction in Monitoring"* of the FRBFE method needs more time resulting from the repeat of the iterative process of the residual extraction than the conventional technique, its small computational time for the order selection highly aids it to establish a fast and time-saving unsupervised learning strategy for feature extraction.

Once the residuals of the AR model at each sensor of the undamaged and damaged conditions have been extracted, it is important to demonstrate their sensitivity to damage. For this purpose, the L_2-norms of the residual vectors $\mathbf{e}_{\mathbf{u}_k}$ and $\mathbf{e}_{\mathbf{d}_z}$ at all sensors are computed. The result is shown in Fig. 5.13, where the norm samples 1–60 belong to the five measurements of Case 1 in the training phase and the remaining 48 samples (61–108) are related to Cases 2–5 of the monitoring period. It is seen that the residual L_2-norms of the damage conditions highly increase in comparison

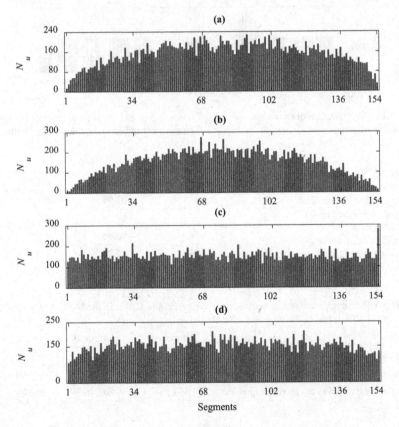

Fig. 5.13 Number of samples (N_u) of $\mathbf{e}_{\mathbf{u}_k}$ based on the segmentation of $\mathbf{e}_{\mathbf{d}_z}$ in the first measurement of Case 1: **a** Sensor 4, **b** Sensor 8, **c** Sensor 11, d Sensor 14

with the corresponding norm values regarding the five normal conditions resulting from the occurrence of damage. These substantial increases confirm the sensitivity of the AR model residuals extracted from the proposed feature extraction method to damage.

5.3.2 Damage Localization by KLDEPM

Utilizing the model order and coefficients obtained from the training phase, the residual vectors e_{u_k} and e_{d_z} at each sensor are extracted to use as the main DSFs of the undamaged and damaged conditions. Determination of the number of samples in each partition of e_{d_z} (N_d) and the number of partitions (c) based on the process of segmentation depends directly on the number of residual sequences, which is equal to $n\text{-}k(k + 1)/2$. Since the total number of acceleration data points (n) is constant, the number of iterations (k) for order selection plays an important role in obtaining different amounts for N_d and c at each sensor. For example, in Fig. 5.12a, the number of iterations (the order of the AR model) at the sensor 9 corresponds to 20. Accordingly, the residual vectors of AR(20) in both the undamaged and damaged states have 23,790 samples by using $n\text{-}k(k + 1)/2$, where $n = 24,000$ and $k = 20$.

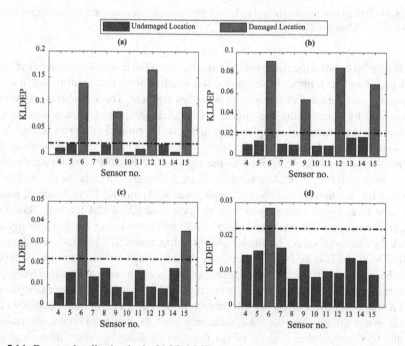

Fig. 5.14 Damage localization in the IASC-ASCE structure by the proposed KLDEPM method: **a** Case 2, **b** Case 3, **c** Case 4, **d** Case 5

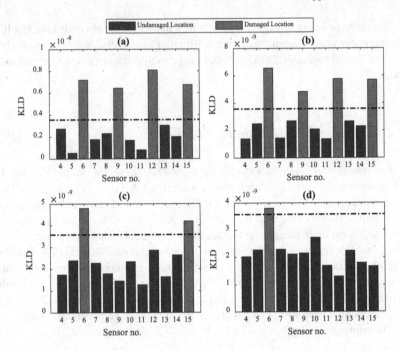

Fig. 5.15 Damage localization in the IASC-ASCE structure by the classical KLD technique: **a** Case 2, **b** Case 3, **c** Case 4, **d** Case 5

Figure 5.14 shows the details of the partitioning of the residual sequences in Cases 2–5 including the number of segments, the number of samples within the first c-1 segments, and the number of samples in the last segment. These amounts are simply obtainable by determining the number of residual samples at each sensor. In the following, the number of samples (N_u) of $\mathbf{e}_{\mathbf{u}_k}$ is determined as shown in Fig. 5.15 for some sensors of the first measurement of Case 1. The obtained amounts of the segmentation process are applied to the equation of KLDEPM to locate damage as illustrated in Fig. 5.16. Additionally, Figs. 5.17 and 5.18 indicate the results of damage localization gained by the conventional KLD and KSTS techniques, which are used to compare their results with the proposed KLDEPM method. Based on the methodology of Monte Carlo (Sect. 4.8.2) and considering the 5% significance level, the threshold values for the KLDEPM, KLD, and KSTS methods are identical to 0.0234, 0.3577, and 0.0461, respectively. As mentioned earlier, the five noise levels have been added to the acceleration responses of Case 1 in order to generate five types of normal conditions; that is, $K = 5$. In this regard, the procedures of feature extraction and damage localization via the information of the generated normal conditions are repeated five times. Using the 5% significance level, the threshold limit is determined by the 95% confidence interval of the five sets of distance values of each of the KLDEPM, KLD, and KSTS methods. Similar to the previous section, the means of

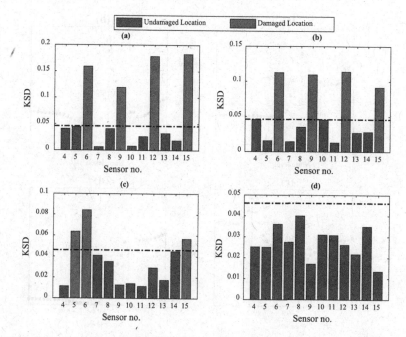

Fig. 5.16 Damage localization in the IASC-ASCE structure by the state-of-the-art KSTS technique: **a** Case 2, **b** Case 3, **c** Case 4, **d** Case 5

distance values between the residual vectors of each damaged state and all normal conditions are calculated to illustrate the result of damage identification.

In Figs. 5.16a, b, 5.17a, b, and 5.18a, b regarding Cases 2 and 3, the damage locations (DL) are precisely identified by the KLDEPM, KLD, and KSTS methods at the east side of the floors 1–4 because the distance values of the sensors 6, 9, 12, and 15 exceed the threshold limits. For Cases 4 and 5 in Figs. 5.16c, d and 5.17c, d, it is seen that both the KLDEPM and KLD methods succeed in locating the damaged areas of the IASC-ASCE structure at the southeast corner of the floors 1 and 4 for the fourth case and the first floor related to the last scenario. These conclusions are based on the distance values of the sensors 6 and 15 in Case 4 and the only sensor 6 for Case 5, all of which are larger the threshold limits of KLDEPM and KLD. Although Fig. 5.18c indicates that KSTS can identify the damage locations at the sensors 6 and 15, a false-positive damage indication (Type I error) is observable at the sensor 5, where it should be indicative of an undamaged location (UDL). Moreover, it can be perceived that this technique is not capable of identifying the damaged area in Case 5 due to a false-negative alarm (Type II error) at the sensor 6 as shown in Fig. 5.18d.

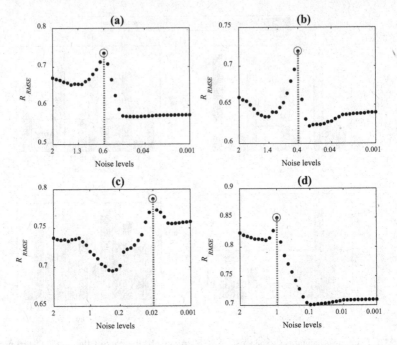

Fig. 5.17 Selection of the optimal noise levels based on the ICEEMDAN method at the sensor 15: **a** Case 1, **b** Case 2, **c** Case 7, **d** Case 9

5.4   Validation of the ICEEMDAN-ARMA, SDC, and MDC Methods by the IASC-ASCE Structure Under Ambient Vibration

In this section, the measured vibration responses of the IASC-ASCE structure acquired from the ambient excitation sources are utilized to validate the accuracy and reliability of some proposed methods. It is attempted to consider both damage patterns of the IASC-ASCE structure for the feature extraction and damage diagnosis procedures. The measured vibration responses of Cases 1, 2, and 4 in the first damage pattern and Cases 7–9 from the second damage pattern, which have previously been listed in Table 5.6 of Sect. 5.3, are utilized. Therefore, one attempts to verify the following methods:

- Feature extraction by the proposed ICEEMDAN-ARMA algorithm (Sect. 3.3).
- Robust order selection by an iterative algorithm (Sect. 2.6.1).
- Optimal IMF extraction by the proposed MPF approach (Sect. 3.3.1).
- Damage diagnosis by the proposed SDC and MDC methods (Sect. 4.5).

Before implementing any process, it is interesting to analyze the acceleration responses of Cases 7–9 to realize whether those are stationary or non-stationary (Table 5.9). Accordingly, Table 5.10 lists the test statistic of the KPSS test (Q_{KPSS}) by

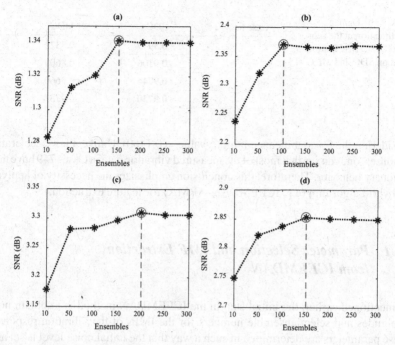

Fig. 5.18 Selection of the appropriate ensemble numbers based on the ICEEMDAN method at the sensor 15: **a** Case 1, **b** Case 2, **c** Case 7, **d** Case 9

Table 5.9 Data analysis in terms of the stationarity vs. non-stationarity by the KPSS hypothesis test in Cases 7–9

Sensor no.	Case no.		
	7	8	9
4	0.7728	1.4116	0.0392
5	34.9770	82.7734	37.0375
6	1.6704	7.4650	0.3135
7	52.2061	52.7082	16.6724
8	23.6519	38.8898	36.4759
9	35.5806	23.3078	8.0341
10	0.0183	0.04872	0.6079
11	2.6696	2.43853	0.6855
12	29.0187	1.3896	1.8804
13	7.6748	6.2906	2.7052
14	21.9142	1.0484	0.3621
15	150.7331	6.2729	0.6314

Table 5.10 Damage
quantification at the sensor 6
of the IASC-ASCE structure
based on SDC and MDC

Case no.	SDC	MDC
2	0.8979	0.8322
4	0.9106	0.8609
8	**0.9744**	0.9660
9	**0.9730**	0.9737

considering the 5% significance limit (c-value $= 0.1460$). Unlike a few acceleration
responses, one can see that most of the measured vibration data in Cases 7–9 have non-
stationary behavior. Therefore, this conclusion emphasizes the necessity of applying
the hybrid algorithm (e.g. ICEEMDAN-ARMA) for feature extraction.

5.4.1 Parameter Selection and IMF Extraction from ICEEMDAN

As mentioned earlier, the initial step in the ICEEMDAN method is to obtain noise
amplitudes and select ensemble numbers for the IMFs of the vibration responses.
These parameters are determined in such a way that the initial noise level is set as 2.
Hence, three sets of noise levels are defined. In the first set, the noise level reduces
from 2 (the initial noise level used in this section) to 0.1 in the step of 0.1; that is,
$(L_n)_1 = 2, 1.9, \ldots, 0.1$. The second set starts with 0.09 and ends with 0.01 in the step
of 0.001, in which case $(L_n)_2 = 0.09, 0.08, \ldots, 0.01$. Eventually, in the third set, the
noise level decreases from 0.009 to 0.001; therefore, $(L_n)_3 = 0.009, 0.008, \ldots, 0.001$.
In this regard, Figs. 5.19 and 5.20 show the optimal values of L_n and the selected
numbers of NE based on the ICEEMDAN method at the sensor 15 of Cases 1 and
2 in the first damage pattern and Cases 7 and 9 for the second pattern. Obtaining
the optimal values of A_n and NE, these amounts are incorporated into the algorithm
of the ICEEMDAN to extract the IMFs. In order to provide a comparative analysis,
Fig. 5.21 illustrates the number of IMFs extracted from the algorithms of EEMD and
ICEEMDAN at all sensors of Cases 1 and 2 in the first damage pattern and Cases 7
and 9 for the second one. It is seen that the ICEEMDAN method reduces the number
of IMFs and avoids extracting redundant modes. Thus, it makes a cost-efficient signal
decomposition process compared to the EEMD technique.

5.4.2 Optimal IMF Extraction from MPF

Once all IMFs from the ICEEMDAN methods have been extracted, the automatic
mode selection approach by using the equation of MPF is utilized to choose IMFs
relevant to damage. Figure 5.22 indicates the IMF energy levels and their participation
factors at the sensor 15 of Cases 1 and 2. In Fig. 5.22b, the second, third, fourth, fifth,

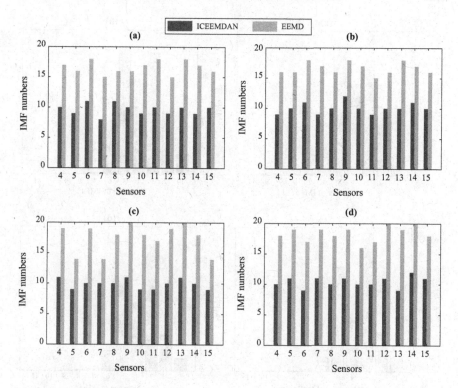

Fig. 5.19 Comparison of the number of IMFs extracted from the EEMD and ICEEMDAN methods: **a** Case 1, **b** Case 2, **c** Case 7, **d** Case 9

and sixth IMFs contain 14%, 25%, 20%, 26, and 14% of participation factors. The MPFs of the second, third, fourth, and fifth IMFs in Fig. 5.22d are identical to 10%, 35%, 26%, and 19%. Considering $\lambda = 90\%$, the above-mentioned IMFs are selected as the optimal time-series modes for ARMA modeling. Figures 5.23 and 5.24 give the optimal numbers of IMFs (n_{opt}) at all sensors of Cases 1, 2, 4, and 7–9.

5.4.3 ARMA Modeling

For ARMA modeling, the selected IMFs are initially arranged in descending order based on their energy levels. Subsequently, the model orders are determined by the iterative order determination method (Sect. 2.6.1) in the healthy states. Figures 5.25 and 5.26 present the orders of ARMA models (p and q) and their p-values associated with the LBQ test in Case 1. The same procedure is carried out to determine the model orders of Case 7. The non-recursive LS technique is applied to estimate the coefficients of ARMA models. It is important to point out that if the number of optimal IMFs in the damaged cases to be more than the corresponding number in

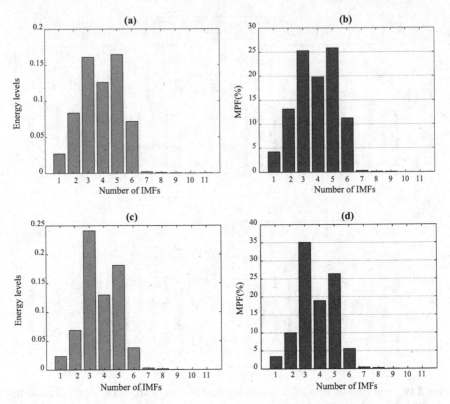

Fig. 5.20 The energy levels of IMFs (*left*) and their participation factors (*right*) at the sensor 15: **a** and **b** Case 1, **c** and **d** Case 2

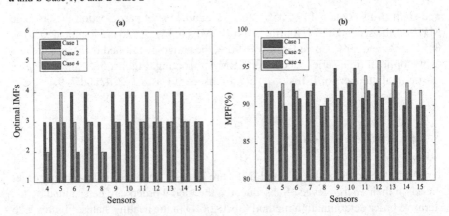

Fig. 5.21 Optimal signal decomposition at all sensors of Cases 1, 2, and 4 **a** the optimal numbers of IMFs (n_{opt}), **b** the mode participation factors

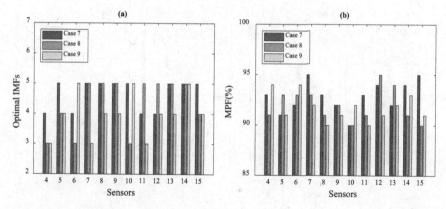

Fig. 5.22 Optimal signal decomposition at all sensors of Cases 7–9 **a** the optimal numbers of IMFs (n_{opt}), **b** the mode participation factors

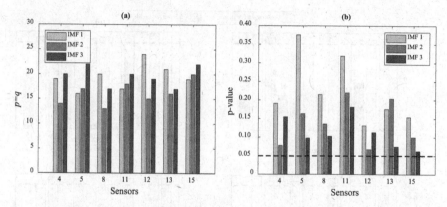

Fig. 5.23 The details of ARMA modeling for sensors with $n_{opt} = 3$ in Case 1 **a** the model orders ($p = q$), **b** the p-values of the ARMA residuals

the undamaged conditions, the additional IMFs are neglected to incorporate them into the process of feature extraction since there are no model specifications (i.e. the model orders and coefficients obtained from the undamaged state) for redundant IMFs of the damaged conditions. For example, the optimal number of IMFs for the undamaged state at the sensor 5 is identical to 3, whereas the corresponding number for Case 2 corresponds to 4. Because there is no ARMA model information for the fourth IMF of this case, it is removed from the process of feature extraction.

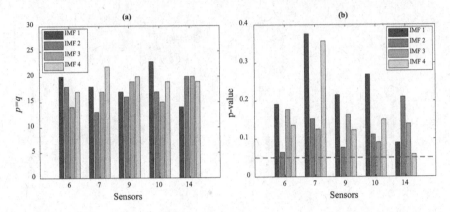

Fig. 5.24 The details of ARMA modeling for sensors with $n_{opt} = 4$ in Case 1 **a** the model orders $(p = q)$, **b** the p-values of the ARMA residuals

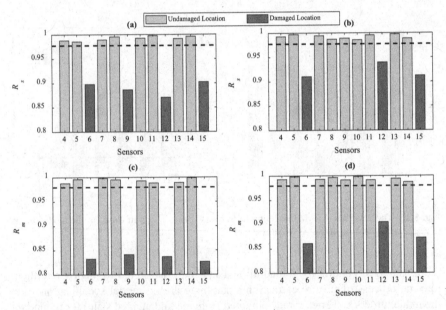

Fig. 5.25 Damage localization in the IASC-ASCE structure in the first damage pattern: **a** Case 2 by SDC, **b** Case 4 by SDC, **c** Case 2 by MDC, **d** Case 4 by MDC

5.4.4 Damage Diagnosis by SDC and MDC

By extracting the model residuals in the healthy (Cases 1 and 7) and damaged (Cases 2, 4, and 8–9) states, the random high-dimensional multivariate sets ($\mathbf{E_u}$ and $\mathbf{E_d}$) are incorporated in the SDC and MDC methods for locating damage in the IASC-ASCE structure. These matrices include 60,000 (the first damage pattern) and 180,000

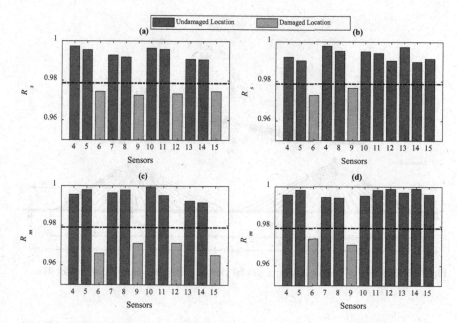

Fig. 5.26 Damage localization in the IASC-ASCE structure in the second damage pattern: **a** Case 8 by SDC, **b** Case 9 by SDC, **c** Case 8 by MDC, **d** Case 9 by MDC

(the second damage pattern) rows (time series samples) and different columns (the number of optimal IMFs, which is different at each sensor and each case).

Before identifying the location of damage, it is essential to define threshold limits for the SDC and MDC methods based on the Monte Carlo simulation technique. For this purpose, 5% random noise is separately added to the vibration responses of Cases 1 and 7. This process is repeated 100 times independently; that is, $K = 100$. Then, the distance correlation values are computed by using the residual sets of the undamaged conditions to obtain two sets of 100 values of d_{SDC} and d_{MDC} for each of the undamaged states. Applying the 95% confidence interval, the threshold values of the first and second damage patterns based on the SDC are equal to 0.9775 and 0.9787, respectively. For the MDC, the thresholds correspond to 0.9795 and 0.9790, respectively. To locate damage, 5% random white noise is also added to the measured vibration responses of each damaged condition. By repeating this procedure 100 times, one can obtain 100 sets of E_u and E_d at each sensor location that lead to 100 sets of d_{SDC} and d_{MDC} values. Eventually, the mean of these collections is calculated to create a vector of distance values, each of which is representative of one of the quantities of d_{SDC} and d_{MDC}. The results of damage localization by the SDC and MDC techniques in the first and second damage patterns are shown in Figs. 5.27 and 5.28, respectively.

In Figs. 5.27a, c, and 5.28a, c for Cases 2 and 8, the damage locations (*DL*) are identified at the east side of the floors 1–4, since the d_{SDC} and d_{MDC} values of the sensors 6, 9, 12, and 15 are less than the threshold limits. Moreover, it can be

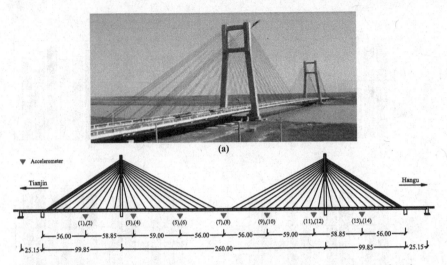

Fig. 5.27 The Tianjin-Yonghe Bridge: **a** general view, **b** main dimensions and sensor locations

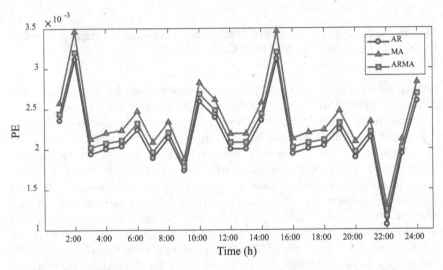

Fig. 5.28 Automatic model identification by the ARMAsel algorithm at the sensor 8 on Day 1 under 24-h vibration measurements

observed in Fig. 5.27b, d that the distance correlation amounts of the sensors 6, 12, and 15 are under the threshold levels implying that the east side of the floors 1, 3, and 4 are the damaged areas in Case 4. Eventually, Fig. 5.28b and 5.28d demonstrate that the first and second floors at the east side of the IASC-ASCE structure are the locations of damage in Case 9 resulting from the d_{SDC} and d_{MDC} quantities of the sensors 6 and 9, which are below the threshold levels. Such observations confirm that

Table 5.11 The undamaged and damaged conditions of the Tianjin-Yonghe Bridge

Day no.	Date	Status	Description
1	January 1	Undamaged	Training phase
2	January 17	Undamaged	
3	February 3	Undamaged	
4	March 19	Undamaged	
5	March 30	Undamaged	
6	April 19	Undamaged	
7	May 5	Undamaged	
8	May 18	Undamaged	
9	July 31	Damaged	Monitoring phase

both the SDC and MDC methods along with the proposed hybrid feature extraction (ICEEMDAN-ARMA) are able to identify the damage locations.

Although both the SDC and MDC methods can accurately identify the damage locations, it needs to evaluate their performance for damage quantification. For this purpose, Table 5.11 presents the results of damage quantification at the sensor 6 in Cases 2 and 4 in the first damage scenario and Cases 8 and 9 for the second pattern based on the SDC and MDC methods. From this figure, one can observe that the severity of damage in Case 2 is higher than the other cases resulting from having the smallest d_{SDC} and d_{MDC} amounts. Moreover, it is seen that these values in Cases 8 and 9 are more than the other cases, which imply their small severities. The comparison of the SDC and MDC values in Table 5.11 presents that MDC is better than SDC due to the inaccurate damage quantification in Cases 8 and 9 by the SDC method.

5.5 Validation of the Automatic Model Identification and PKLD-MSD Methods by the SMC Cable-Stayed Bridge

The SMC benchmark problem is a cable-stayed bridge (the Tianjin-Yonghe Bridge) as shown in Fig. 5.29a. This is one of the earliest cable-stayed bridges constructed in Mainland China. The Tianjin-Yonghe Bridge involves a main span of 260 m and two side spans of 25.15 and 99.85 m as depicted in Fig. 5.29b. The bridge is 510 m long and 11 m wide including 9 m for vehicles and 2 × 1 m for pedestrians. The concrete towers, connected by two transverse beams, are 60.5 m tall. More details of the bridge can be found in Li et al. (2014).

The Tianjin-Yonghe Bridge was opened to traffic since December 1987; however, after 19 years of operation in 2005, some serious cracks were found at the bottom of a girder segment over the mid-span. Furthermore, some stay cables, particularly those near the anchors, were severely corroded. Applying a sophisticated SHM system organized by the Center of SMC at the Harbin Institute of Technology in China, the

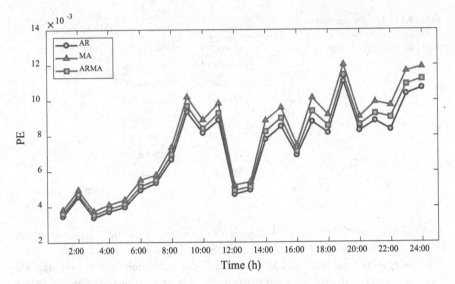

Fig. 5.29 Automatic model identification by the ARMAsel algorithm at the sensor 8 on Day 9 under 24-h vibration measurements

bridge was monitored in 2007 after a major rehabilitation program for replacing the damaged girder segment and all the stay cables between 2005 and 2007. In August 2008, new damage patterns during a routine inspection of the bridge were found in the girders. Acceleration time histories from January to August 2008 during 12 days were measured by 14 single-axis accelerometers, as shown in Fig. 5.29b, under operational (traffic loads) and/or environmental (temperature and wind) variability conditions. The accelerations include 24 data sets of 1 h with the sampling frequency of 100 Hz and a time interval of 0.01 s consisting of 360,000 data samples.

In this section, the acceleration responses of 13 sensors including the accelerometers 1–9 and 11–14 on January 1, January 17, February 3, March 19, March 30, April 19, May 5, May 18, and July 31 regarding the second damage problem defined in Li et al. (2014) are utilized to verify the following approaches:

- Automatic model identification approach (Sect. 2.5.2)
- Optimal and robust model order selection by an improved two-stage algorithm for big data (Sect. 2.6.3)
- Early damage detection by the proposed PKLD-MSD method (Sect. 4.6.2)

Note that the data belonging to the sensor 10 include meaningless measurement samples. Therefore, the time series data of this sensor is not considered. Furthermore, since the first 8 days (January 1 - May 18) of measurements are representative of the undamaged states (Nguyen et al. 2014), these are utilized in the training phase; that is, $n_L = 8$. In this regard, the last day of measurements on July 31 is considered during the

Table 5.12 Evaluation of the overfitting problem for the AR orders at 15:00 of Days 1, 3, 5, and 7

Sensor no.	Day no.							
	1		3		5		7	
	R^2	$Adj\text{-}R^2$	R^2	$Adj\text{-}R^2$	R^2	$Adj\text{-}R^2$	R^2	$Adj\text{-}R^2$
1	0.9205	0.9069	0.9051	0.8942	0.9175	0.8962	0.9375	0.9111
2	0.9205	0.9069	0.9051	0.8942	0.9175	0.8962	0.9375	0.9111
3	0.9329	0.9195	0.8967	0.8798	0.9019	0.8801	0.9080	0.8962
4	0.9132	0.8944	0.9040	0.8933	0.9043	0.8884	0.9409	0.9236
5	0.9277	0.9012	0.9363	0.9120	0.9219	0.9044	0.8949	0.8814
6	0.8827	0.8747	0.9060	0.8938	0.9112	0.8929	0.9179	0.8822
7	0.9153	0.9002	0.9239	0.9071	0.9024	0.8990	0.9092	0.8931
8	0.9222	0.9063	0.8937	0.8761	0.9422	0.9269	0.8770	0.8689
9	0.9003	0.8899	0.9074	0.8870	0.8967	0.8800	0.9112	0.8843
11	0.9318	0.9133	0.9221	0.8959	0.9044	0.8845	0.8639	0.8511
12	0.9041	0.8879	0.9114	0.8971	0.8930	0.8782	0.8948	0.8756
13	0.9292	0.9124	0.9216	0.8975	0.9006	0.8888	0.8892	0.8735
14	0.9223	0.9085	0.8768	0.8702	0.9267	0.9025	0.9037	0.8821

inspection period. Therefore, it is attempted to discern whether the proposed PKLD-MSD method is able to detect early damage under the environmental variability and distinguish the normal condition of the bridge from the damaged one.

Table 5.12 presents the healthy and damaged conditions of the bridge within 9 days to use in the baseline and inspection phases. It is worth mentioning that the structural states on Days 1–8 in this table are equivalent to $S_{N_1}, S_{N_2}, \ldots, S_{N_L}$, in which the structural conditions on Day 1 and Day 8 are designated by S_{N_1} and S_{N_L}, respectively.

5.5.1 Automatic Model Identification

The process of model identification is carried out by engineering and statistical aspects. From an engineering viewpoint, the ambient excitations (i.e. wind, traffic, etc.) applied to the bridge are unmeasurable and unknown implying the unavailability of input data. As explained in Sect. 2.4, one can utilize the ARMA or ARARX representations for modeling the vibration responses. From a statistical aspect, the main difference between these models lies in the fact that ARARX is consistent with the AR process, whereas ARMA conforms to both the AR and MA processes (Box et al. 2015; Ljung 1999). Using the automatic model identification approach based on the ARMAsel algorithm, the best single AR, MA, and ARMA models are initially obtained from the statistical criteria CIC and GIC. In the following, the PE values of these three representations are calculated to choose the most proper time series model between ARMA and ARARX. As samples, Figs. 5.30 and 5.31 show the results of automatic model identification at the sensor 8 of Days 1 and 9.

It is clear that the PE values of the best AR representations in all test measurements are smaller than the other best models. This conclusion confirms that the acceleration responses conform to the AR process. Therefore, the ARARX model is identified as the most appropriate time series model based on the statistical and engineering aspects to utilize in the process of feature extraction. Note that the same conclusion can be achieved at the other sensors and days. Figure 5.32 shows the comparison of computational time between the automatic model identification method and the Box-Jenkins methodology. For this comparative analysis, the acceleration response of the sensor 8 in the 8th test measurement of Day 1 is used and decomposed into nine data sets consisting of 6000, 10,000, 18,000, 30,000, 60,000, 120,000, 180,000, 240,000, and 360,000 samples, respectively. Figure 5.32 presents noticeable information about the performance of the graphical and numerical methods. Regardless of the requirement and limitation of the Box-Jenkins methodology, which depends on the user expertise, it can be observed that it provides a better performance concerning the computational time than the ARMAsel algorithm in the small data sets (i.e. the first two sets). However, it is realized that the automatic model identification method needs a smaller time (e.g. less than 60 s for the last data set with 3,600,000 samples) compared to the Box-Jenkins methodology after the third data set. This means that the proposed ARMAsel algorithm is highly superior to the Box-Jenkins methodology when the dimensionality of data sequences increases.

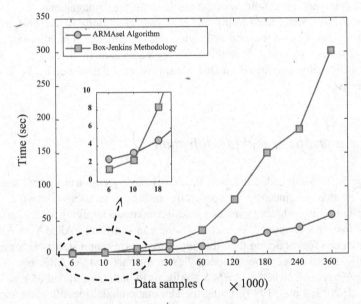

Fig. 5.30 Comparison of the computational time between the automatic model identification method (ARMAsel algorithm) and Box-Jenkins methodology

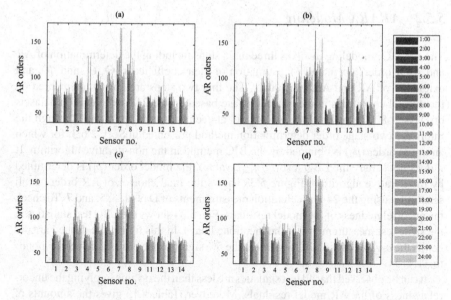

Fig. 5.31 Determination of AR orders at all sensors and all test measurements: **a** Day 1, **b** Day 3, **c** Day 5, and **d** Day 7

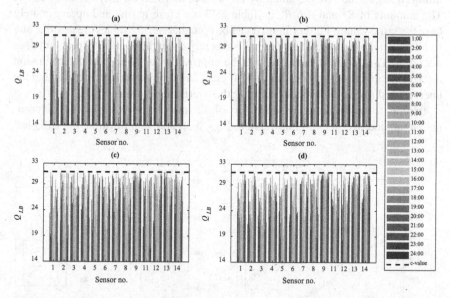

Fig. 5.32 Residual analysis by the LBQ test statistics in all test measurements: **a** Day 1, **b** Day 3, **c** Day 5, and **d** Day 7

5.5.2 ARARX Modeling

The ARARX modeling involves three main steps including the determination of AR and ARX orders ($p, \bar{p}$, and $\bar{r}$), estimation of their coefficients (Θ, $\bar{\Theta}$, and $\bar{\Phi}$), and extraction of AR and ARX residuals. The first two steps are only implemented on the normal conditions in the training phase based on the 24-h acceleration datasets of Days 1–8. The order of the AR model at each sensor is determined through the improved two-stage order determination method presented in Sect. 2.6.3, for which an initial order (p_0) is obtained by the BIC method in the non-iterative algorithm. If it does not satisfy the LBQ test ($Q_{LB} \geq$ c-value), the model order (p_i) is determined by the iterative algorithm. Figure 5.33 shows the final choices of AR orders at all sensors by using the 24-h acceleration measurements of Days 1, 3, 5, and 7. To check the uncorrelatedness of the model residuals, Fig. 5.34 shows the LBQ test statistics in the same test measurements of the mentioned days. In this figure, the dashed arrows depict the c-value of the LBQ test under the 5% significance limit, which corresponds to 31.4104.

It can be observed that all test statistics are less than the c-value implying the uncorrelatedness of the AR model residuals. Moreover, Table 5.13 gives the amounts of R-squared and adjusted R-squared statistics to investigate the probability of overfitting by using the orders gained by all sensors at 15:00 on Days 1, 3, 5, and 7. The amounts of R^2 and Adj-R^2 in Table 5.13 are close to one and approximately have the same values. Furthermore, it can be seen that Adj-R^2 values are positive and smaller than R^2 quantities. These conclusions prove that the overfitting problem does not occur. Thus, one can conclude that the improved two-stage order determination method is capable of determining the sufficient and optimal orders by extracting the uncorrelated residuals without occurring the overfitting problem.

In order to demonstrate the superiority of the improved two-stage order determination method over the original version, Figs. 5.35 and 5.36 illustrate the LBQ test

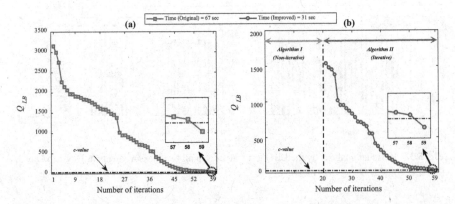

Fig. 5.33 Comparison of the computational time for determining the minimum AR order ($p = 59$) of the acceleration time histories of the sensor 9 at 17:00 on Day 1: **a** the original method presented in Sect. 2.6.1, **b** the improved method presented in Sect. 2.6.3

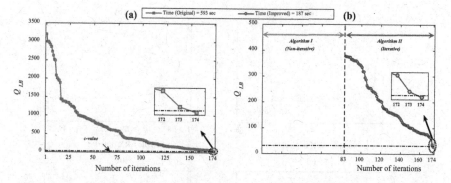

Fig. 5.34 Comparison of the computational time for determining the maximum AR order ($p = 174$) of the acceleration time histories of the sensor 7 at 11:00 on Day 1: **a** the original method presented in Sect. 2.6.1, **b** the improved method presented in Sect. 2.6.3

Table 5.13 Different cases of classifying residual datasets based on the 24-h vibration measurements

Case no.	Days 1–8		Day 9	
	Training	Testing	Testing	Testing
1	1:12	13:24	1:12	13:24
2	1:2:24	2:2:24	1:2:24	2:2:24
3	13:24	1:12	13:24	1:12
4	2:2:24	1:2:24	2:2:24	1:2:24

statistics and computational time needed for determining the minimum and maximum AR orders on Day 1 (i.e. $p_{min} = 59$ and $p_{max} = 174$). In Figs. 5.35a and 5.36a, the iterative process begins with one, whereas the improved method firstly computes the initial orders by the BIC method and checks the correlation of the model residuals. As can be seen in Figs. 5.35b and 5.36b, the initial orders are identical to 20 and 83, respectively. Since these values fail in satisfying the LBQ test, the final orders are determined by the second algorithm. The results in these figures indicate that both methods are able to determine the adequate orders of AR models by extracting the uncorrelated residuals so that $Q_{LB} <$ c-value at the p_{min} and p_{max}. Although the improved method consists of the two stages (i.e. the non-iterative and iterative algorithms) and assesses both the uncorrelatedness of the model residuals and overfitting problem, it needs less computational time in comparison with the original technique. Therefore, it can be concluded that the improved two-stage method provides a more cost-efficient order determination algorithm compared to the original technique for the SHM of large-scale structures with the high-dimensional data.

To obtain the orders of ARX representation associated with the second part of the ARARX model, it is important to point out that the sum of ARX orders should be smaller than the AR order; that is, $\bar{p} + \bar{r} \leq p$ (Ljung 1999). This means that any selection of ARX orders, which their sum to be less than p, is plausible. Accordingly,

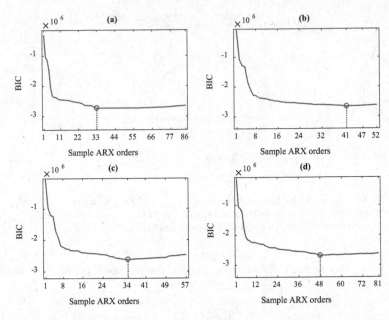

Fig. 5.35 Optimal ARX orders ($\bar{p} = \bar{r}$) of the sensor 8 at 20:00: **a** Day 1, **b** Day 3, **c** Day 5, and **d** Day 7

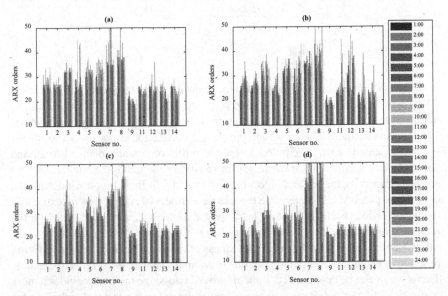

Fig. 5.36 Optimal ARX orders obtained from the BIC method at all sensors and all test measurements: **a** Day 1, **b** Day 3, **c** Day 5, and **d** Day 7

there are many acceptable ARX orders. In such circumstances, an efficient way is to use one of the information criteria to select optimal amounts among all possible ARX orders (Wei 2006). For this purpose, one initially assumes that $\bar{p} = \bar{r}$. By using the BIC method for the final choice, it is possible to examine different orders from one to the half of the obtained AR order (i.e. $\bar{p} = \bar{r} = 1, 2, \ldots, p/2$) and choose an optimal amount with the smallest BIC quantity. As an instance, Fig. 5.37 shows the orders of ARX models gained by the BIC method at the sensor 8 and 20:00 on Days 1, 3, 5, and 7. Additionally, Fig. 5.38 illustrates the optimal ARX orders at all sensors of these days.

5.5.3 The Conventional RBFE Approach

Given the sufficient and optimal AR orders obtained from the normal conditions in the training phase, the model coefficients on Days 1–9 are initially estimated by the least-squares technique. Using such information, the residual sets of AR models for the structural conditions in both the baseline and inspection phases are extracted to utilize in the ARX representations as the input datasets. In the following, the ARX residuals of the structural conditions on Days 1–8 are extracted by using the obtained model orders ($\bar{p}$ and $\bar{r}$) and estimated model coefficients ($\bar{\Theta} = [\bar{\theta}_1 \ldots \bar{\theta}_{\bar{p}}]$ and $\bar{\Phi} = [\bar{\varphi}_1 \ldots \bar{\varphi}_{\bar{r}}]$) as the main DSFs of the normal conditions in the training phase. The same procedure is carried out to extract the residuals of the ARX models for the structural state on Day 9, which are used as the main DSFs of the current state of the structure. It needs to mention that the orders and coefficients of the ARX representations regarding the normal conditions should be allocated to the structural state in the inspection phase. Since the number of measurements and sensors in the structural states of Days 1–9 are similar, each residual set of these days makes a three-dimensional matrix ($360,000 \times 13 \times 24$) containing 360,000 data points obtained from 13 sensors and 24-h measurements. According to the RBFE method, it is necessary to use the ARARX orders and coefficients obtained from the undamaged states on Days 1–8 in order to extract the residual data sets regarding the current state. Therefore, eight residual sets related to the structural state on Day 9 are extracted as the DSFs of the current condition.

5.5.4 Early Damage Detection by PKLD-MSD

To detect early damage by the proposed PKLD-MSD methodology, it is initially necessary to define the training and testing data sets as explained in Sect. 4.6. These are obtained by measuring the similarity between the reference ($\mathbf{E_u}$) and current ($\mathbf{E_d}$) residual vectors through the PKLD method. At first, it is essential to specify the number of samples and partitions of the current residual vector after its arrangement in ascending order. Considering 360,000 residual sequences (n), the number of samples

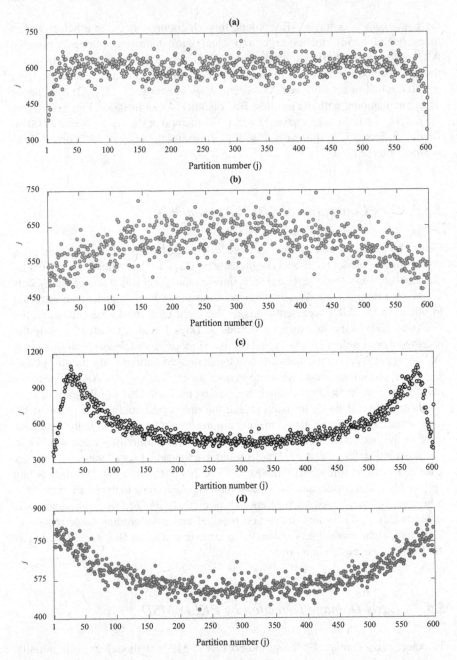

Fig. 5.37 Partitioning of the reference residual vectors of the sensor 7 at 15:00: **a** Day 1, **b** Day 3, **c** Day 5, **d** Day 7

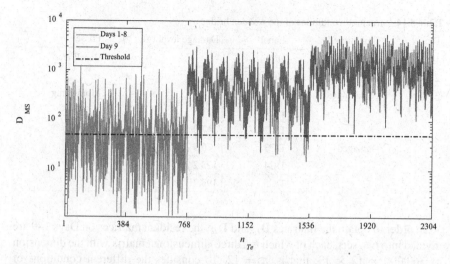

Fig. 5.38 Early damage detection by the proposed PKLD-MSD method in Case 1

$(N_d = \sqrt{n})$ in each partition of the arranged $\mathbf{E_d}$ corresponds to 600. Using these amounts, the number of partitions $(c = n/N_d)$ becomes 600 as well. It needs to mention that the correction factor (γ_c) is identical to zero resulting from the equality of the number of samples and partitions. This means that the last partition of $\mathbf{E_d}$ has the same residual sequences as the other partitions. In the following, the reference residual vector is partitioned by the maximum entropy approach without any arrangement. Given 600 partitions, Fig. 5.39 displays the number of samples for each partition of $\mathbf{E_u}$ at the sensor 7 on Days 1, 3, 5, and 7 (the 8th test measurement).

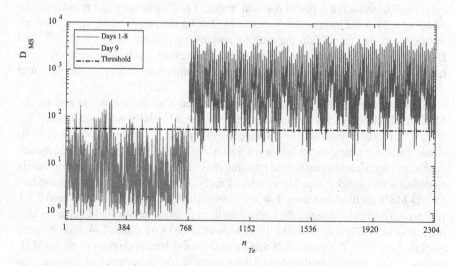

Fig. 5.39 Early damage detection by the proposed PKLD-MSD method in Case 2

Table 5.14 Structural conditions of the wooden bridge

Day	Condition	Label	Damage level (g)	Phase
1	Undamaged	HC1	–	Baseline
2	Undamaged	HC2	–	
3	Undamaged	HC3	–	Monitoring
		DC1	23.5	
		DC2	47.0	
3	Damaged	DC3	70.5	
		DC4	123.2	
		DC5	193.7	

In order to obtain the matrices $\mathbf{D}_{Tr}$ and $\mathbf{D}_{Te}$, the residual matrices on Days 1–9 are divided into two sets, each of which is a three-dimensional matrix with the dimension of 360,000 × 12 × 13; that is, $n_T = 12$. To consider the different conditions of environmental effects, four cases of divisions are assigned as presented in Table 5.14.

For example, in Case 1, one supposes that the residual sets associated with the first 12-h vibration measurements on Days 1–8 are used to determine $\mathbf{D}_{Tr}$ based on the consideration of 50% of DSFs of the normal conditions in the training process. By contrast, it is assumed that the residual datasets regarding the second 12-h measurements on Days 1–8 as well as the two sets of 12-h measurements of Day 9 are utilized to determine $\mathbf{D}_{Te}$. As another example in Case 2, one assumes that the residual datasets of the odd hours of 24-h measurements of Days 1–8 are incorporated in the training phase to obtain $\mathbf{D}_{Tr}$. In this case, the even hours of these days along with the two sets of the odd and even hours of measurements on Day 9 are utilized in the inspection phase to make the testing matrix. Note that since the testing dataset in each case is comprised of the three types of residual matrices and there are eight normal conditions ($n_L = 8$) in the training phase, $n_U = 24$. Taking the values of n_S, n_T, n_L, and n_U into account, one can obtain the training and testing datasets as $\mathbf{D}_{Tr} \in \mathbb{R}^{13 \times 768}$ and $\mathbf{D}_{Te} \in \mathbb{R}^{13 \times 2304}$, where the first 768 samples of the testing matrix belong to the structural states on Days 1–8 and the remaining 1536 samples are related to the structural condition on Day 9.

Applying the training dataset, the unsupervised learning model trained by the MSD method comprises the mean vector $\bar{\mathbf{d}}_{Tr} \in \mathbb{R}^{13}$ and covariance matrix $\Sigma_{Tr} \in \mathbb{R}^{13 \times 13}$. Novelty detection for damage identification is based on computing MSD values by driving the column vectors of the testing dataset through the trained model. A similar way is also implemented by using the column vectors of the training matrix to define a threshold value. The results of early damage detection by the proposed PKLD-MSD method in Cases 1–4 are shown in Figs. 5.40, 5.41, 5.42, and 5.43, respectively. In these figures, the horizontal lines display the threshold values, which are identical to 50.5460, 55.4107, 49.3532, and 51.6819 for Cases 1–4. It is observed in Figs. 5.40, 5.41, 5.42, and 5.43 that there are considerable deviations of the MSD values on Day 9 from the threshold values implying the occurrence of damage. The

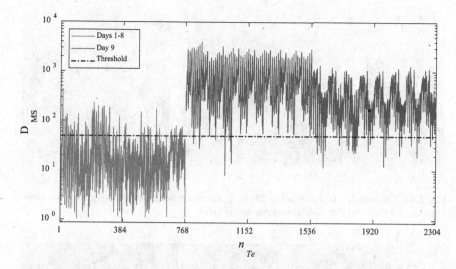

Fig. 5.40 Early damage detection by the proposed PKLD-MSD method in Case 3

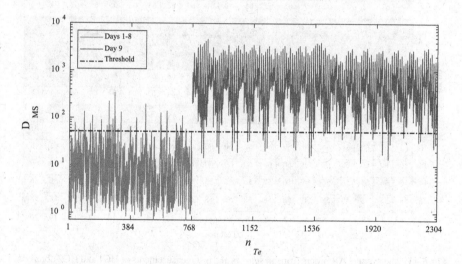

Fig. 5.41 Early damage detection by the proposed PKLD-MSD method in Case 4

false acceptance (Type II) errors in Cases 1–4 correspond to 1.8%, 4.4%, 3.2%, and 2.5%, respectively. Due to the very low Type II errors in the MSD quantities on Day 9 (i.e. 1536 samples of $\mathbf{D}_{Te}$ from 768 to 2304), one can deduce that the proposed PKLD-MSD method and the methodology for the estimation of threshold limits succeed in detecting damage.

On the other hand, when the samples of training data do not broadly cover the environmental effects, it can be seen irregular variations in Figs. 5.40 and 5.41 regarding

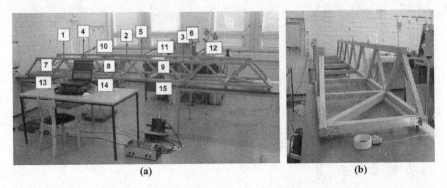

Fig. 5.42 The wooden bridge (Kullaa 2011): **a** the experimental setup along with the sensor numbers and measurement directions, **b** the lateral view

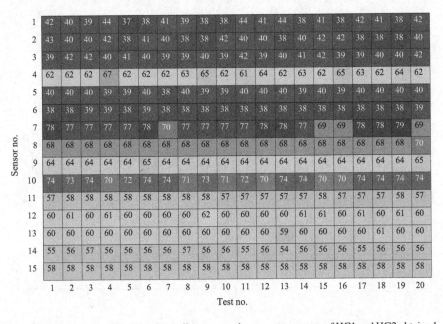

Sensor no. \ Test no.	1	2	3	4	5	6	7	8	9	10	11	12	13	14	15	16	17	18	19	20
1	42	40	39	44	37	38	41	39	38	38	44	41	44	38	41	38	42	41	38	42
2	43	40	40	42	38	41	40	38	38	42	40	40	38	40	42	42	38	38	38	40
3	39	42	42	40	41	40	39	39	40	39	42	39	40	41	42	39	39	40	40	42
4	62	62	62	67	62	62	62	63	65	62	61	64	62	63	62	65	63	62	64	62
5	40	40	40	39	39	40	38	40	39	39	40	40	40	39	40	39	40	40	40	40
6	38	38	39	39	38	39	38	38	38	38	38	38	38	38	38	38	38	38	38	39
7	78	77	77	77	77	78	70	77	77	77	77	78	78	77	69	69	78	78	79	69
8	68	68	68	68	68	68	68	68	68	68	68	68	68	68	68	68	68	68	68	70
9	64	64	64	64	64	65	64	64	64	64	64	64	64	64	64	64	64	64	64	65
10	74	73	74	70	72	74	74	71	73	71	72	70	74	74	70	70	74	74	74	74
11	57	58	58	58	58	58	58	58	58	57	57	57	57	57	58	57	57	57	58	57
12	60	61	60	61	60	60	60	60	62	60	60	60	60	61	61	60	61	60	61	60
13	60	60	60	60	60	60	60	60	60	60	60	60	59	60	60	60	60	61	60	60
14	55	56	57	56	56	56	56	57	56	56	55	56	54	56	56	56	55	56	56	56
15	58	58	58	58	58	58	58	58	58	58	58	58	58	58	58	58	58	58	58	58

Fig. 5.43 The average AR orders from all sensors and test measurements of HC1 and HC2 obtained from the iterative approach presented in Sect. 2.6.1

the MSD quantities of Day 9 for Cases 1 and 3. Furthermore, the false negative (Type I) errors in Cases 1-4 are identical to 46.7%, 6.2%, 14.2%, and 5.7%, respectively. Unlike Cases 2 and 4, the high Type I errors in Cases 1 and 3 are most likely caused by the inappropriate consideration of the environmental variability by choosing the residual sets of the only first or second 12-h vibration measurements on Days 1–8 for obtaining the training dataset (Fig. 5.44).

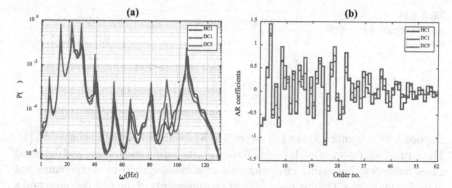

Fig. 5.44 Comparison of the outputs of AR modeling at Sensor 4 for the first test measurement of HC1, DC1, and DC5: **a** AR spectra, **b** AR coefficients

5.6 Validation of Spectral-Based and Multi-level Distance-Based Methods by the Wooden Truss Bridge

The wooden truss bridge is a benchmark structure subjected to actual environmental variability conditions in the laboratory (Kullaa 2011). The truss was constructed by wood as shown in Fig. 5.45. It was equipped with 15 uniaxial sensors for measuring acceleration time histories at three different longitudinal positions. Figure 5.45a shows the experimental setup, the sensor numbers, and the measurement directions. An electro-dynamic shaker with a random forced excitation source was applied to vibrate the bridge in the vertical, transverse, and torsional modes. The acceleration

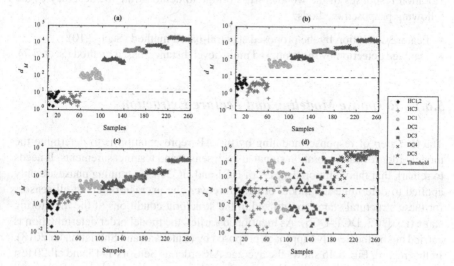

Fig. 5.45 Damage detection by the proposed multi-level distance-based methodology and AR spectrum: **a** Scenario 1, **b** Scenario 2, **c** Scenario 3, **d** Scenario 4

Table 5.15 The different types of sensor deployment scenarios

Sensor deployment scenarios	Sensor labels
1	2, 4, 6, 7, 9, 11, 15
2	1, 3, 5, 12,14
3	4, 8, 13
4	14

responses were comprised of 8192 samples at 32 s with sampling frequency 256 Hz. The measurements were made during three days including undamaged and damaged cases under varying actual environmental conditions caused by temperature and humidity variations. On this basis, all test measurements of the first two days and a few measurements of the third day present the normal conditions of the truss structure (Kullaa 2016).

Damage was simulated by adding a mass point at the end of the girder close to the sensor 4 on the third day. The values of mass were 23.5, 47.0, 70.5, 123.2, and 193.7 g, each of which reflects a level of damage severity. Table 5.15 summarizes and classifies the undamaged and damaged cases of the wooden bridge.

It needs to mention that the number of test measurements n_t is equal to 20 for the undamaged and damaged states. On the other hand, according to the common procedure of machine learning, the first two undamaged states of the structure (i.e. HC1, HC2) are incorporated in the baseline phase. In this case, the number of normal conditions n_c is 2. Furthermore, the structural states HC3 and DC1–5 are utilized in the monitoring stage as the current states. Therefore, it is attempted to realize whether the proposed methods are able to detect HC3 and DC1–DC5 as the normal and damaged conditions, respectively. This section focuses on using the measured vibration responses of the wooden truss bridge to demonstrate the accuracy of the following proposed methods:

- Feature extraction by the proposed spectral-based method (Sect. 2.10).
- Damage detection by the proposed multi-level distance-based method (Sect. 4.7).

5.6.1 Response Modeling and Feature Extraction

The first step of response modeling by the AR representation is to determine the model order for each vibration signal at each sensor and test measurements. It needs to remark that the vibration datasets of HC1 and HC2 in the training phase are only applied to determine the model order. Moreover, the average orders at all sensors for these structural states are utilized in the structural conditions of the monitoring stage (i.e. HC3, DC1–DC5). As mentioned earlier, the model order determination is carried out by the iterative approach proposed by (Entezami and Shariatmadar 2018). In this regard, Fig. 5.46 shows the average AR order as Sensors 1–15 and all 20 test measurements.

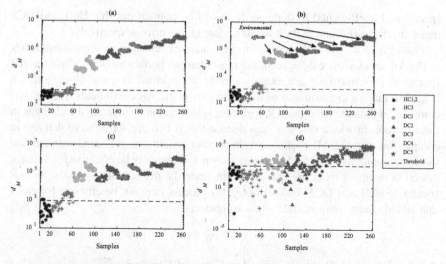

Fig. 5.46 Damage detection by the proposed multi-level distance-based methodology without the AANN algorithm: **a** Scenario 1, **b** Scenario 2, **c** Scenario 3, **d** Scenario 4

Using the obtained average orders, the AR spectrum at each sensor is estimated by the Burg's method all sensors, and test measurements of all structural states. In this regard, Fig. 5.47 illustrates the comparison between the AR spectra and coefficients at Sensor 4 for the first test measurement of HC1, DC1, and DC5. The main objective of this comparison is to indicate the sensitivity of these different features (i.e. AR

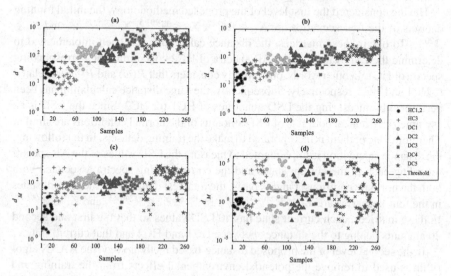

Fig. 5.47 Damage detection by the proposed multi-level distance-based methodology using the PSD: **a** Scenario 1, **b** Scenario 2, **c** Scenario 3, **d** Scenario 4

spectra and coefficients) to damage. It should be pointed out that DC1 and DC5 represent the lowest and highest levels of damage severity, respectively.

From Fig. 5.47a, it is apparent that there are clear changes in some frequencies of the AR spectra. As can be seen, the most changes pertain to DC5, while the AR spectra of HC1 and DC1 are roughly similar. By contrast, one can observe that it is hard to make a consequence of the variations in the AR coefficients. Hence, this comparison indicates that the AR spectrum is able to better indicate the changes in the structure, in which case one can deduce that it is more sensitive to damage in comparison with the AR coefficients. In another conclusion, it is demonstrated that the direct comparison of the AR spectra may not sufficiently be effective for damage detection resulting from the inability to indicate the discrepancy between the AR spectra of HC1 and DC1. Furthermore, this process may not be efficient for cases that include many sensors and test measurements.

5.6.2 Damage Detection Under Limited Sensors and Environmental Effects

To detect damage in the Wooden Bridge by considering the limited number of sensors, some scenarios are defined to simulate this condition. For this purpose, one assumes that a few sensors were mounted on the structure. In this regard, four types of sensor deployment scenarios are incorporated as listed in Table 5.15. For each scenario, the AR spectrum of that sensor is only applied to detect damage based on the proposed multi-level distance-based methodology.

Having considered the first level of the proposed methodology, the initial training datasets of the four scenarios are $\mathbf{X}_1 \in \mathbb{R}^{20\times7}$, $\mathbf{X}_2 \in \mathbb{R}^{20\times5}$, $\mathbf{X}_3 \in \mathbb{R}^{20\times3}$, and $\mathbf{X}_4 \in \mathbb{R}^{20\times1}$. To obtain these matrices, the distance calculation has been implemented to determine the LSD values of the AR spectra of HC2 concerning the corresponding spectra of HC1. In other words, one initially considers that $P(\omega)$ and $\bar{P}(\omega)$ are related to HC1 and HC2, respectively. Subsequently, the same distance calculation has been performed by measuring the LSD quantities of HC1 for HC2. Since the LSD is an asymmetric distance, both sets of LSD amounts of the normal conditions are similar. Therefore, one of them is incorporated to make the training datasets. In the following, the distance calculation is implemented to measure the LSD values of the AR spectra regarding each of the current states and the corresponding spectra associated with both the normal conditions. On this basis, the total test datasets for the current states in the four scenarios are $\mathbf{Z}_1 \in \mathbb{R}^{240\times7}$, $\mathbf{Z}_2 \in \mathbb{R}^{240\times5}$, $\mathbf{Z}_3 \in \mathbb{R}^{240\times3}$, and $\mathbf{Z}_4 \in \mathbb{R}^{240\times1}$. In these matrices, each current state has 40 LSD values so that the first and second 20 amounts belong to the distances between HC1 and HC2 and that current state.

In the second level of the proposed distance-based methodology, the AANN algorithm is used to remove the potential environmental effects from the training and test datasets. According to Kramer's recommendations (Figueiredo et al. 2011), the neural networks for each of the scenarios have 10 nodes in each mapping and de-mapping layer and 2 nodes in the bottleneck layer. Using the trained networks, the new normalized training and test matrices $\mathbf{E_x}$ and $\mathbf{E_z}$ are determined to apply to the MSD for early damage detection as shown in Fig. 5.46, where the dashed lines depict the threshold limits gained by the 95% confidence interval of the MSD values from the only training data.

From Fig. 5.46a, b, it is seen that DC1–DC5 are reasonably detected as the damaged conditions because their d_M values exceed the thresholds. Additionally, one can observe that these distance amounts are split appropriately so that DC5 (the highest level of damage severity) and DC1 (the lowest level of damage severity) have the largest and smallest MSD values, and these quantities increase by increasing the damage severity level from DC1 to DC5. As the other conclusion, it is clear that the d_M values of HC3, which has been considered as one of the current states in the monitoring phase, are similar to the corresponding quantities of HC1 and HC2. Approximately, the same conclusions can be observed in Fig. 5.46a but not as good as the first and second scenarios. Although all d_M values of DC1-DC5 exceed the threshold limits, the MSD quantities of DC1 and DC2 are the same. Nonetheless, Fig. 5.46d shows unreliable and inaccurate damage detection in the fourth scenario despite the appropriate damage detection in DC4 and DC5. This observation indicates that the use of one sensor may not be effective and useful for damage detection.

The previous results of damage detection are based on using the AANN algorithm for removing the potential environmental effects from the training and test datasets. As a comparative work, it is attempted to evaluate the influence of this algorithm. Accordingly, the process of damage detection is implemented by using the initial training and test matrices without applying the AANN algorithm. Figure 5.47 indicates the results related to this situation in all four scenarios. As Fig. 5.47a–c appears, there are discrepancies between the first and second d_M values for each current state. For example, these dissimilarities are highlighted in Fig. 5.47b. Despite reasonable detection of DC1–DC5 as the damaged conditions, it is seen that the second MSD quantities of HC3 exceed the threshold limits indicating false alarms. Approximately, the same conclusions are observable in Fig. 5.47d; however, the result of damage detection is not acceptable.

5.7   Conclusions

In this chapter of the book, all improved and proposed methods for the model identi-fication, order selection, feature extraction, and damage diagnosis have been verified by several experimental structures. Briefly, the results obtained from these struc-tures have demonstrated that the proposed and improved feature extraction methods provide reliable DSFs for damage detection, localization, and quantification. In time series modeling, the proposed order determination methods have indicated that they are able to yield adequate and accurate orders of time-invariant linear models. These orders have enabled the time series representations to generate uncorrelated residuals, which are indicative of the adequacy and accuracy of time series modeling. Moreover, it has been observed that the proposed feature classification methods and proposed distance measures succeed in detecting, locating, and quantifying the damage. The summary and main conclusions of these methods will be mentioned in the next chapter of this book.

References

Box GEP, Jenkins GM, Reinsel GC, Ljung GM (2015) Time series analysis: forecasting and control, 5th edn. Wiley

Dyke SJ, Bernal D, Beck J, Ventura C (2003) Experimental phase II of the structural health monitoring benchmark problem. In: Proceedings of the 16th ASCE engineering mechanics conference, 2003

Entezami A, Shariatmadar H (2018) An unsupervised learning approach by novel damage indices in structural health monitoring for damage localization and quantification. Struct Health Monit 17(2):325–345

Figueiredo E, Park G, Farrar CR, Worden K, Figueiras J (2011) Machine learning algorithms for damage detection under operational and environmental variability. Struct Health Monit 10(6):559–572

Figueiredo E, Park G, Figueiras J, Farrar C, Worden K (2009) Structural health monitoring algorithm comparisons using standard data sets. Los Alamos National Laboratory: LA-14393

Figueiredo E, Todd M, Farrar C, Flynn E (2010) Autoregressive modeling with state-space embedding vectors for damage detection under operational variability. Int J Eng Sci 48(10):822–834

Kullaa J (2011) Distinguishing between sensor fault, structural damage, and environmental or operational effects in structural health monitoring. Mech Syst Signal Process 25(8):2976–2989

Kullaa J (2016) Experimental data of the wooden bridge: http://users.metropolia.fi/~kullj/

Leybourne SJ, McCabe BPM (1994) A consistent test for a unit root. J Bus Econ Stat 12(2):157–166

Li S, Li H, Liu Y, Lan C, Zhou W, Ou J (2014) SMC structural health monitoring benchmark problem using monitored data from an actual cable-stayed bridge. Struct Control Health Monit 21(2):156–172

Ljung L (1999) System identification: theory for the user, 2nd edn. Prentice-Hall, Upper Saddle River, NJ

Nguyen T, Chan T, Thambiratnam D (2014) Controlled Monte Carlo data generation for statistical damage identification employing Mahalanobis squared distance. Struct Health Monit 13(4):461–472

Wei WWS (2006) Time series analysis: univariate and multivariate methods, 2nd edn. Pearson Addison Wesley

Chapter 6
Summary and Conclusions

6.1 Introduction

This research work has focused on the health monitoring of civil structures based on the statistical pattern recognition paradigm. The process of feature extraction and statistical decision-making have been defined as the main steps of this paradigm for the structural health monitoring (SHM) and damage diagnosis, which includes the damage detection, localization, and quantification.

For the process of feature extraction, this research work has considered time series modeling and time–frequency signal decomposition techniques. In time series modeling, the major challenges have included selecting an appropriate time series representation, determining robust and optimal orders of time series models, developing residual-based feature extraction (RBFE) algorithms for dealing with some limitations such as the effects of environmental and/or operational variability (EOV) and high-dimensional data.

Since no order determination and parameter estimation are needed to implement an RBFE algorithm in the monitoring phase, it is superior to the coefficient-based feature extraction (CBFE) approach. However, one needs to improve the conventional RBFE algorithm in an effort to cope with some limitations and challenges associated with the big data and the effects of EOV. Hence, three RBFE methods have been proposed to extract the residuals of time series models as the main DSFs. These methods have been the developed residual-based feature extraction (DRBFE), and the fast residual-based feature extraction (FRBFE).

Feature extraction under non-stationary vibration responses caused by unmeasurable and unknown ambient vibration is another important issue in SHM. In the case of using non-stationary vibration signals for feature extraction, this work has proposed a hybrid algorithm as the combination of signal decomposition technique and time series model, which has been called ICEEMDAN-ARMA. An automatic approach by mode participation factor (MPF) has been proposed to choose the relevant IMFs to damage.

Apart from the process of feature extraction, it is important to propose or introduce robust and innovative distance-based novelty detection methods for damage diagnosis by high-dimensional features under the EOV. These methods have been classified as the univariate distance measures such as Kullback–Leibler divergence with empirical probability measure (KLDEPM), parametric assurance criterion (PAC), and residual reliability criterion (RRC) and multivariate distance measures including sample distance correlation (SDC) and modified distance correlation (MDC). The univariate distance methods have been presented to locate damage, while the multivariate approaches have been introduced to detect damage. Furthermore, a hybrid method for novelty detection as the combination of partition-based Kullback–Leibler divergence (PKLD) and MSD, which has been called PKLD-MSD, has been proposed to detect damage using random high-dimensional features in the presence of EOV conditions. Threshold estimation is a vital step in a novelty detection method. An inappropriate estimate may lead to Type I and Type II errors.

Finally, several experimental benchmark structures have been considered to validate the effectiveness and performance of the above-mentioned methods along with several comparative studies. The experimental structures have been the four-story laboratory frame of the Los Alamos National Laboratory, the wooden bridge, the four-story steel structure of the IASC-ASCE SHM problem in the second phase, and the Tianjin-Yonghe bridge in China.

The proceeding parts of this chapter of the book are as follows: Sect. 6.2 mentions the main conclusions related to the process of feature extraction. In Sect. 6.3, the main conclusions of statistical decision-making for damage diagnosis are drawn. Finally, Sect. 6.4 presents some suggestions for further developments.

6.2 Main Conclusions of Feature Extraction

Based on the results of the experimental structures, the first conclusion is that the implementation of initial data analysis for assessing the measured vibration responses in terms of the stationarity vs. non-stationarity and linearity vs. non-linearity is necessary before choosing a feature extraction approach. When the vibration responses are stationary and linear, the use of time-invariant linear models such as AR, ARMA, and ARARX is plausible. In the case of using the non-stationary data, the proposed hybrid algorithms are very suitable.

Concerning the procedure of model identification, the proposed automatic ARMAsel (ARMA Selection) algorithm has facilitated the selection of a time series model without the user inference and expertise. The comparison of this algorithm with the Box-Jenkins methodology (the graphical approach) has shown that the ARMAsel algorithm is more cost-efficient and faster than the Box-Jenkins methodology, particularly in the use of high-dimensional time series datasets.

In relation to the RBFE, the proposed DRBFE method has enhanced the conventional RBFE technique for obtaining uncorrelated residuals. Moreover, the comparison between the proposed FRBFE and conventional RBFE methods has revealed that the former is faster than the latter in the way that the process of order selection had the highest influence on the computational time of feature extraction. Hence, it is preferable to use the FRBFE method in the presence of high-dimensional data.

In the case of using the non-stationary vibration signals after the initial data analysis, it is recommended to utilize the proposed hybrid algorithm ICEEMDAN-ARMA. It has been observed that this method is highly suitable for feature extraction under non-stationary signals. When the nature of vibration measurements caused by ambient vibration is unknown, it is highly preferable to apply these approaches. The comparative work has revealed that the proposed ICEEMDAN method reduces the number of redundant IMFs compared to the well-known EEMD technique. This merit leads to a cost-efficient signal decomposition process and prevents obtaining useless information. The proposed automatic IMF selection method aids the ICEEMDAN method in decreasing the redundant modes and choosing the most relevant IMFs to damage based on their energy levels and participation factors. Since this approach provides an automatic algorithm, it can overcome the limitation of using the visual or experience criteria by the user. The results of signal decomposition and IMF selection procedures have shown that the IMFs extracted from the ICEEMDAN and MPF approaches with the high energy levels are more suitable for damage detection.

Regarding the proposed spectral-based method, the results have shown that the AR spectrum is a reliable feature for damage detection using the limited sensors. The comparison between the AR spectrum and coefficient has revealed that the former is able to better indicate the changes in the structure, in which case one can deduce that it is more sensitive to damage in comparison with the AR coefficients. Moreover, it has been observed that the direct use of the AR spectra is not sufficiently effective for damage detection. Therefore, it is important to apply the multi-level distance method for this problem.

6.3 Main Conclusions of Damage Diagnosis

The first step of damage diagnosis is to detect damage. For this purpose, the PKLD-MSD method has yielded accurate results. The great advantage of this method is that it can deal with the limitation of using the random high-dimensional features for early damage detection. The proposed PKLD-MSD method could successfully detect early damage in the cable-stayed bridge using the random high-dimensional features and in the presence of EOV conditions. It has been seen that the use of a threshold highly increases the reliability of damage detection. However, one needs to utilize a large number of training samples in order to incorporate all possible EOV in normal conditions.

The second step of damage diagnosis is to locate damage. Most of the distance measures including PAC, RRC, KLDEMP, SDC, and MDC have been proposed to identify the location of the damage. The important conclusion is that the accuracy of damage localization depends strongly on the location of the sensor. This means that the location of damage should be close to the sensor. This conclusion indicates the importance of sensor placement in SHM and damage localization.

It has been seen that both the PAC and RRC methods are able to simultaneously locate and quantify damage even without estimating the threshold limit. The results of damage localization via these methods have demonstrated that the EOV conditions do not have any effects on the results of damage diagnosis.

Both the SDC and MDC methods have been able to precisely identify the locations of single and multiple damage cases with different severity levels by using high-dimensional multivariate residual sets. However, it has been seen that the SDC method could not give reliable and accurate results of damage quantification, whereas the MDC method has been correctly capable of estimating the level of damage severity. The assessment of the effects of the sampling frequency and time duration on the performances of the SDC and MDC methods have revealed that the increases in sampling frequency and time duration improve the reliability of damage localization results and the use of insufficient information in data acquisition leads to the serious false positive and false negative errors.

The proposed KLDEPM method has provided the accurate results of damage localization by dealing with the limitations of using the random high-dimensional time series features and the effects of the EOV conditions. For the problem of damage quantification, it has been observed that the smallest and largest KLDEPM amounts at the damaged area, by considering all damaged conditions, are representative of the lowest and highest damage severities, respectively. This method has been better than the classical KSTS for locating small damage cases. The comparison between the KLDEPM and KLD techniques has shown that the former is faster than the latter and requires a short time for decision-making. Therefore, it is highly suitable for damage diagnosis under high-dimensional features in terms of the rate of decision-making.

Regarding the proposed multi-level distance method, the results have demonstrated that this method is highly able to detect early damage under the limited sensors and environmental effects. However, the use of one sensor has not been succeeded in providing accurate and reasonable results. It has been seen that the AANN algorithm enhances the performance of the proposed multi-level methodology and removes the effects of environmental variability. Finally, it has been observed that the proposed multi-level distance method and AR spectra outperform the traditional MSD in conjunction with the AR coefficients for damage detection under limited sensors.

6.4 Further Developments

Despite the accurate results of damage detection, localization, and quantification via the proposed methods, it is necessary to state some suggestions for further works.

In most cases, the final decision about the occurrence of damage in the first step of damage diagnosis (i.e. early damage detection) is based on finding or observing any deviation from threshold limits (Sarmadi and Karamodin 2020; Deraemaeker and

Worden 2018; Entezami et al. 2020a; Entezami et al. 2020b; Entezami et al. 2020c; Sarmadi et al. 2020; Sarmadi and Entezami 2020). This is an important issue because thresholds are obtained by statistical properties of the outputs of the distance-based novelty detection methods in the normal conditions. The simplest approach is to use the standard confidence interval (e.g. 95% confidence interval) based on the central limit theorem and the Gaussian distribution assumption. However, this approach may fail in providing accurate threshold limits when the distribution of the outputs of novelty detection methods or distance measures in the normal conditions is non-Gaussian (Sarmadi and Karamodin 2020). For further works, it is recommended to use the other kinds of threshold limit determination via the extreme value theorem and Bootstrap methods.

In this research work, the AR, ARMA, and ARARX models have been widely used in the process of feature extraction. Such time series representations are useful for conditions that the vibration responses are linear and stationary. For the non-stationary measurements, although this research work has proposed a hybrid algorithm, it is suggested to evaluate and utilize time-varying linear models such as Time-Varying Autoregressive (TV-AR) (Avendano-Valencia and Fassois 2014), Time-Varying Autoregressive Moving Average (TV-ARMA) (Poulimenos and Fassois 2009), Autoregressive Integrated Moving Average (ARIMA) (Omenzetter and Brownjohn 2006), etc. for feature extraction without the aid of the time–frequency signal decomposition techniques.

The accuracy of damage localization depends strongly on the location of the sensor (Capellari et al. 2016, 2017; Bruggi and Mariani 2013). Therefore, it is suggested to pay more attention to the issue of sensor placement for the problem of damage localization.

References

Avendano-Valencia L, Fassois S (2014) Stationary and non-stationary random vibration modelling and analysis for an operating wind turbine. Mech Syst Sig Process 47(1):263–285

Bruggi M, Mariani S (2013) Optimization of sensor placement to detect damage in flexible plates. Eng Optim 45(6):659–676

Capellari G, Chatzi E, Mariani S (2016) Optimal sensor placement through Bayesian experimental design: effect of measurement noise and number of sensors. In: Multidisciplinary digital publishing institute proceedings, vol 2, p 41

Capellari G, Chatzi E, Mariani S (2017) Cost-benefit optimization of sensor networks for SHM applications. In: Multidisciplinary digital publishing institute proceedings. vol 3, p 132

Deraemaeker A, Worden K (2018) A comparison of linear approaches to filter out environmental effects in structural health monitoring. Mech Syst Sig Process 105:1–15

Entezami A, Sarmadi H, Salar M, De Michele C, Arslan AN (2020a) A novel data-driven method for structural health monitoring under ambient vibration and high-dimensional features by robust multidimensional scaling. Struct Health Monit, In press

Entezami A, Shariatmadar H, Mariani S (2020) Early damage assessment in large-scale structures by innovative statistical pattern recognition methods based on time series modeling and novelty detection. Adv Eng Softw 150:102923

Entezami A, Shariatmadar H, Sarmadi H (2020) Condition assessment of civil structures for structural health monitoring using supervised learning classification methods. Iranian J Sci Technol, Trans Civ Eng 44(1):51–66

Omenzetter P, Brownjohn JMW (2006) Application of time series analysis for bridge monitoring. Smart Mater Struct 15(1):129

Poulimenos AG, Fassois SD (2009) Output-only stochastic identification of a time-varying structure via functional series TARMA models. Mech Syst Sig Process 23(4):1180–1204

Sarmadi H, Entezami A (2020) Application of supervised learning to validation of damage detection. Arch Appl Mech, In press

Sarmadi H, Entezami A, Saeedi Razavi B, Yuen K-V (2020) Ensemble learning-based structural health monitoring by Mahalanobis distance metrics. Struct Control Health Monit:e2663

Sarmadi H, Karamodin A (2020) A novel anomaly detection method based on adaptive Mahalanobis-squared distance and one-class kNN rule for structural health monitoring under environmental effects. Mech Syst Sig Process 140:106495

Printed in the United States
by Baker & Taylor Publisher Services

Printed in the United States
By Bookmasters